T0339770

Ulam Stability of Operators

Mathematical Analysis and its Applications Series

Ulam Stability of Operators

Authors

Janusz Brzdęk
Pedagogical University, Department of Mathematics,
Podchorążych 2, 30-084 Kraków, Poland

Dorian Popa
Technical University of Cluj-Napoca, Department of
Mathematics, 28 Memorandumului Street, 400114,
Cluj-Napoca, Romania

Ioan Raşa
Technical University of Cluj-Napoca, Department of
Mathematics, 28 Memorandumului Street, 400114,
Cluj-Napoca, Romania

Bing Xu
Sichuan University, Department of Mathematics,
No. 29 Wangjiang Road, 610064, Chengdu, China

Series Editor
Themistocles M. Rassias

Academic Press is an imprint of Elsevier
125 London Wall, London EC2Y 5AS, United Kingdom
525 B Street, Suite 1800, San Diego, CA 92101-4495, United States
50 Hampshire Street, 5th Floor, Cambridge, MA 02139, United States
The Boulevard, Langford Lane, Kidlington, Oxford OX5 1GB, United Kingdom

Copyright © 2018 Elsevier Inc. All rights reserved.

No part of this publication may be reproduced or transmitted in any form or by any means, electronic
or mechanical, including photocopying, recording, or any information storage and retrieval system,
without permission in writing from the publisher. Details on how to seek permission, further information
about the Publisher's permissions policies and our arrangements with organizations such as the
Copyright Clearance Center and the Copyright Licensing Agency, can be found at our website:
www.elsevier.com/permissions.

This book and the individual contributions contained in it are protected under copyright by the Publisher
(other than as may be noted herein).

Notices
Knowledge and best practice in this field are constantly changing. As new research and experience
broaden our understanding, changes in research methods, professional practices, or medical treatment
may become necessary.

Practitioners and researchers must always rely on their own experience and knowledge in evaluating and
using any information, methods, compounds, or experiments described herein. In using such information
or methods they should be mindful of their own safety and the safety of others, including parties for
whom they have a professional responsibility.

To the fullest extent of the law, neither the Publisher nor the authors, contributors, or editors, assume any
liability for any injury and/or damage to persons or property as a matter of products liability, negligence
or otherwise, or from any use or operation of any methods, products, instructions, or ideas contained in
the material herein.

British Library Cataloguing in Publication Data
A catalogue record for this book is available from the British Library

Library of Congress Cataloging-in-Publication Data
A catalog record for this book is available from the Library of Congress

ISBN: 978-0-12-809829-5

For information on all Academic Press publications
visit our website at https://www.elsevier.com/books-and-journals

Working together
to grow libraries in
developing countries

www.elsevier.com • www.bookaid.org

Publisher: Candice Janco
Acquisition Editor: Graham Nisbet
Editorial Project Manager: Susan Ikeda
Production Project Manager: Surya Narayanan Jayachandran
Designer: Matthew Limbert

Typeset by SPi Global, India

We dedicate this monograph to Professor Themistocles M. Rassias, the editor of this series of books, on the occasion of the 40th anniversary of the publication of his first paper on the stability of functional equations, which together with his numerous other papers strongly influenced the development of the theory of Ulam stability.

Janusz Brzdęk, Dorian Popa, Ioan Raşa, Bing Xu

CONTENTS

ACKNOWLEDGMENT

We thank the Elsevier staff for guidance throughout the publishing process; we especially thank Susan E. Ikeda and her team. We are also grateful to all the anonymous referees for carefully reviewing and improving our preliminary proposal of this monograph.

Janusz Brzdęk, Dorian Popa, Ioan Raşa, Bing Xu
June 30, 2017

PREFACE

The aim of this book is not to present a survey of various papers dealing with the Ulam stability. We would not be able to do this in one book, because this area of research is too vast at the moment. Moreover, there are several such books published already and we do not want to copy their approach. Rather, we try to propose a somewhat new systematic approach to investigating Ulam stability. Therefore, after presenting some general results we show numerous examples of applications in various forms of difference, differential, functional and integral equations. Certainly, we use various previously published results, but in this book they are very often extended, generalized, and/or modified. So, it can be said that this book contains numerous outcomes that are new and unpublished, so far.

In this way we would like to show possible directions for future research and thus stimulate further investigations of Ulam stability as well as other related areas of mathematics. For this reason we do not tend to obtain the most general version of outcomes and further possible generalizations of them can be easily visible in many cases.

In the first chapter we present a brief introduction to the subject and cite several somewhat randomly selected results, providing references to sources with more detailed information on Ulam stability.

Our book presents, for the first time in unified and systematic way, some novel approaches to Ulam stability of numerous, mainly linear, operators. Moreover, it has a unique position of presenting up-to-date knowledge on subjects that have been treated only marginally in other similar books published. It includes, in particular, a lot of information on stability of several difference equations, functional equations in a single variable, various types of differential equations, and some integral equations.

It collects and compares suitable results from papers that have been published several years ago but also those published very recently; also, it unifies, complements, generalizes, and updates that information. Whenever it is suitable, open problems have been stated that suggest further possible exploration in the corresponding areas. The book is of interest to specialized researchers in the fields of various types of analysis, operator theory, difference and functional equations and inequalities, and differential and integral equations.

Janusz Brzdęk, Dorian Popa, Ioan Raşa, Bing Xu
June 30, 2017

ABOUT THE AUTHORS

Janusz Brzdęk is Professor of Mathematics at Pedagogical University of Cracow (Poland). He has published numerous papers on Ulam's type stability (e.g., of difference, differential, functional, and integral equations), its applications, and connections to other areas of mathematics; he has been editor of several books and special volumes focused on such subjects. He was also the chairman of the organizing and scientific committees of several international conferences on Ulam's type stability and functional equations and inequalities.

Dorian Popa is Professor of Mathematics at Technical University of Cluj-Napoca (Romania). He is the author of numerous papers on Ulam's type stability of functional equations, differential equations, linear differential operators, and positive linear operators in approximation theory. His other papers deal with the connections of Ulam's type stability with some topics pertaining to multivalued analysis (e.g., the existence of a selection of a multivalued operator satisfying a functional inclusion associated with a functional equation).

Ioan Raşa is Professor of Mathematics at Technical University of Cluj-Napoca (Romania). He has published papers on Ulam's type stability of differential operators and several types of positive linear operators arising in approximation theory. He is author/co-author of many papers connecting Ulam's stability with other areas of mathematics (functional analysis, approximation theory, and differential equations). He is a co-author (with. F. Altomare et al.) of the book Markov Operators, Positive Semigroups and Approximation Processes, de Gruyter, 2014.

Bing Xu is Professor of Mathematics at Sichuan University (China). She has published many papers on Ulam's type stability (e.g., of difference, differential, functional, and integral equations), its applications and connections to iterative equations, and multivalued analysis. Xu is co-author with W. Zhang et al. of the book Ordinary Differential Equations, Higher Education Press, 2014.

CHAPTER 1

Introduction to Ulam stability theory

Contents

Abstract

We describe the origin of Ulam stability theory, methods, and approaches, as well as some relevant results on this topic. In particular, we mention the preliminary result of G. Pólya and G. Szegö (published in 1925), describe the problem of S.M. Ulam (1909-1984), posed in 1940, and the partial solution to it that was published in 1941 by D.H. Hyers. Next, we present the further analogous outcomes of Ulam and Hyers (e.g., those published in 1945, 1947, 1952) and the results of T. Aoki (1950), D.G. Bourgin (1949, 1951), Th.M. Rassias (1978), J. Rätz (1980), P. Găvruţa (1994) and others. We then discuss the stability results for various equations (difference, differential, functional, and integral) providing suitable examples of them. We also depict the notions of superstability and hyperstability, and we present some remarks on the notion of nonstability.

1. Historical background

The stability problem of functional equations was originally raised by Stanisław Marcin Ulam (cf. [73, 144]) in the fall term of the year 1940, when he gave a wide ranging talk before the Mathematics Club of the University of Wisconsin, discussing a number of unsolved problems. Among these was the following question concerning the approximate homomorphisms of groups:

Ulam Stability of Operators
http://dx.doi.org/10.1016/B978-0-12-809829-5.50001-5
Copyright © 2018 Elsevier Inc. All rights reserved.

We are given a group G_1 and a metric group G_2 with metric d. Given $\varepsilon > 0$, does there exist a $\delta > 0$ such that if $f : G_1 \to G_2$ satisfies

$$d(f(xy), f(x)f(y)) \leq \delta, \qquad x, y \in G_1,$$

then a homomorphism $g : G_1 \to G_2$ exists with

$$d(f(x), g(x)) \leq \varepsilon, \qquad x \in G_1?$$

However, a somewhat similar problem was considered earlier by G. Pólya and G. Szegö [119], for $G_1 = \mathbb{N}$ (positive integers) and $G_2 = \mathbb{R}$ (reals). They have obtained the following result.

Theorem 1. *Suppose that a sequence $(a_n)_{n \in \mathbb{N}}$ of real numbers satisfies*

$$a_m + a_n - 1 < a_{m+n} < a_m + a_n + 1, \qquad m, n \in \mathbb{N}.$$

Then the limit

$$\omega = \lim_{n \to \infty} \frac{a_n}{n}$$

exists and satisfies

$$\omega n - 1 < a_n < \omega n + 1, \qquad n \in \mathbb{N}.$$

The first partial answer to Ulam's question came within a year, when D.H. Hyers [73] proved a result that can be stated as follows:

Theorem 2. *Let E_1 and E_2 be Banach spaces and let $f : E_1 \to E_2$ be a transformation such that, for some $\delta > 0$,*

$$\|f(x + y) - f(x) - f(y)\| \leq \delta, \qquad x, y \in E_1.$$

Then the limit

$$g(x) = \lim_{n \to \infty} \frac{f(2^n x)}{2^n}$$

exists for each $x \in E_1$ and $g : E_1 \to E_2$ is the unique additive transformation satisfying

$$\|f(x) - g(x)\| \leq \delta, \qquad x \in E_1.$$

Moreover, if f is continuous at least in one point $x \in E_1$, then g is continuous everywhere in E_1.

Furthermore, if the function $\mathbb{R} \ni t \to f(tx)$ is continuous for each $x \in E_1$, then g is linear.

So if G_1, G_2 are the additive groups of Banach spaces, this theorem provides a positive answer to Ulam's question with $\varepsilon = \delta$. Shortly, we describe that result stating that the additive Cauchy equation,

$$f(x + y) = f(x) + f(y),$$

is *Hyers-Ulam stable* (or *has the Hyers-Ulam stability*) in the class of functions $f : E_1 \to E_2$.

Below, we describe the method of proof used in [73]; we call it the *direct method*; for information and further references concerning other methods see [32, 36, 74].

It is easy to prove, by induction, that

$$\|2^{-n} f(x) - f(2^{-n} x)\| \leq \delta(1 - 2^{-n}), \qquad x \in E_1, n \in \mathbb{N}.$$

Write

$$g_n(x) = \frac{f(2^n x)}{2^n}, \qquad x \in E_1, n \in \mathbb{N} \cup \{0\}.$$

Then

$$\|g_m(x) - g_n(x)\| \leq \delta \frac{1 - 2^{m-n}}{2^m}, \qquad x \in E_1, \ m, n \in \mathbb{N} \cup \{0\}, n > m. \qquad (1.1)$$

Hence $(g_n(x))_{n \in \mathbb{N}}$ is a Cauchy sequence for each $x \in E_1$, and since E_2 is complete, there exists the limit function $g : E_1 \to E_2$,

$$g(x) = \lim_{n \to \infty} g_n(x), \qquad x \in E_1.$$

Clearly,

$$\|g_n(x + y) - g_n(x) - g_n(y)\| \leq \frac{\delta}{2^n}, \qquad x, y \in E_1, n \in \mathbb{N},$$

whence letting $n \to \infty$ we obtain the additivity of g. Next, (1.1) with $m = 0$, yields

$$\|f(x) - g(x)\| \leq \delta, \qquad x \in E_1. \qquad (1.2)$$

Note that, for any additive $\tilde{g} : E_1 \to E_2$, satisfying the inequality

$$\|f(x) - \tilde{g}(x)\| \leq \delta, \qquad x \in E_1,$$

we have

$$\|g(x) - \tilde{g}(x)\| \leq 2\delta, \qquad x \in E_1,$$

which implies that $g = \tilde{g}$.

Clearly, if f is continuous at point $y \in E_1$, then from (1.2) we deduce that g is bounded in a neighborhood of y, and consequently g is continuous.

Finally, if for a fixed x, the function $f_x : \mathbb{R} \ni t \to f(tx)$ is continuous, then the function $\mathbb{R} \ni t \to g(tx)$ is additive and bounded on any finite interval, whence continuous and therefore linear. So, the assumption of continuity of f_x for each $x \in E_1$ implies the linearity of g.

2. Stability of additive mapping

T. Aoki [10] extended the result of Hyers by considering the case where the Cauchy difference $f(x + y) - f(x) - f(y)$ is not necessarily bounded. He proved the following:

Theorem 3. *Let E_1 and E_2 be Banach spaces, and $f : E_1 \to E_2$ be such that*

$$\|f(x + y) - f(x) - f(y)\| \le K(\|x\|^p + \|y\|^p), \qquad x, y \in E_1,$$

with some $K \ge 0$ and $p \in [0, 1)$. Then there exists a unique additive $g : E_1 \to E_2$ such that

$$\|f(x) - g(x)\| \le \frac{2K}{2 - 2^p}\|x\|^p, \qquad x \in E_1.$$

In 1951 D.G. Bourgin [24] provided a further generalization in which he simply stated (without a proof) that, for every function f mapping a normed space X into a Banach space Y and fulfilling the inequality

$$\|f(x + y) - f(x) - f(y)\| \le \Phi(x, y), \qquad x, y \in X, \tag{1.3}$$

with a function $\Phi : X \times X \to [0, \infty)$ that satisfies the condition

$$\sum_{i=1}^{\infty} 2^{-i}\Phi(2^{i-1}x, 2^{i-1}x) < \infty, \qquad x \in X,$$

and certain additional assumptions, there exists an additive transformation $g : X \to Y$ with

$$\|f(x) - g(x)\| \le \sum_{i=1}^{\infty} 2^{-i}\Phi(2^{i-1}x, 2^{i-1}x), \qquad x \in X.$$

Th.M. Rassias in [125] considered a particular case of that stability result with $\Phi(x, y) = K(\|x\|^p + \|y\|^p)$, for $K \ge 0$ and $p \in [0, 1)$, but under the additional assumption of the continuity of the function $\mathbb{R} \ni t \to f(tx)$ for each fixed $x \in E_1$. He obtained in this way not only the additivity, but also the linearity of g. In view of a large influence of [125] and numerous other Rassias papers (see, e.g., [122, 127, 128, 131]), on the further study of stability problem for functional equations, the stability phenomenon of such type has been very often called the *Hyers-Ulam-Rassias stability*.

It is easily seen that the method of proof of Theorem 2 works also in the case where $(E_1, +)$ is "only" a semigroup. But J. Rätz [132] noticed even more (see also [13]). Namely, let $(G, *)$ be a power-associative groupoid, i.e., G is a non-empty set endowed with a binary operation $* : G \times G \to G$ such that the left powers satisfy

$$x^{m+n} = x^m * x^n, \qquad m, n \in \mathbb{N}, \; x \in G.$$

Let $(Y, |\cdot|)$ be a topological vector space over the field $\mathbb{Q}$ of rationals, with $\mathbb{Q}$ topologized by its usual absolute value $|\cdot|$. The stability results [132, Theorems 4 and 5] have been obtained by a direct method and can be stated as follows.

Theorem 4. *Let V be a non-empty bounded $\mathbb{Q}$-convex subset of Y, containing the origin, and assume that Y is sequentially complete. Let $f : G \to Y$ satisfy the following two conditions*

$$f((x * y)^{k^n}) = f(x^{k^n} * y^{k^n}), \qquad x, y \in G,$$

for some integer $k \geq 2$ and all $n \in \mathbb{N}$, and

$$f(x) + f(y) - f(x * y) \in V, \qquad x, y \in G.$$

Then there exists a function $g : G \to Y$ such that

$$g(x) + g(y) = g(x * y), \qquad x, y \in G,$$

and

$$f(x) - g(x) \in \bar{V}, \qquad x \in G,$$

where $\bar{V}$ is the sequential closure of V. When Y is a Hausdorff space, g is uniquely determined.

Let us finally mention one more stability result proved by P. Găvruţa [68] for the Cauchy equation (with inequality (1.3)); it can be stated as follows.

Theorem 5. *Suppose that $(X, +)$ is an abelian group, Y is a Banach space, and $\Phi : X \times X \to [0, \infty)$ is a mapping such that*

$$\tilde{\Phi}(x, y) := \sum_{i=0}^{\infty} 2^{-i} \Phi(2^i x, 2^i y) < \infty, \qquad x, y \in X.$$

If $f : X \to Y$ satisfies

$$\|f(x + y) - f(x) - f(y)\| \leq \Phi(x, y), \qquad x, y \in X,$$

then there exists a unique mapping $g : X \to Y$ such that

$$g(x + y) = g(x) + g(y), \qquad x, y \in X,$$

and

$$\|f(x) - g(x)\| \le \frac{1}{2}\tilde{\Phi}(x, x), \qquad x \in X.$$

Besides the above results a great number of papers on stability have been published, generalizing and extending Ulam's problem in various directions and to other equations or operators (see, e.g., [2, 14, 26, 32, 33, 34, 38, 39, 63, 70, 75, 81, 86, 97, 129, 140, 146]), in particular to various conditional versions of the homomorphism equation (see, e.g., [36, 45]). For some discussions on suitable terminology we refer to [108, 109, 110, 111].

3. Approximate isometries

Let (E_1, d_{E_1}) and (E_2, d_{E_2}) be metric spaces. A mapping $I : E_1 \to E_2$ is called an isometry if I satisfies the equation

$$d_{E_2}(I(x), I(y)) = d_{E_1}(x, y), \qquad x, y \in E_1.$$

D.H. Hyers and S.M. Ulam [76] proved the following stability result for isometries between real Hilbert spaces.

Theorem 6. *Let E be a complete real Hilbert space. Let $\varepsilon > 0$ and T be a surjection of E into itself that is an ε-isometry, that is,*

$$|\rho(T(x), T(y)) - \rho(x, y)| < \varepsilon, \qquad x, y \in E,$$

where ρ denotes the inner product in E. Assume that $T(0) = 0$. Then the limit

$$I(x) = \lim_{n \to \infty} \frac{T(2^n x)}{2^n}$$

exists for every $x \in E$ and the transformation I is a surjective isometry of E into itself, which satisfies

$$\|T(x) - I(x)\| < 10\varepsilon, \qquad x \in E.$$

This result of Hyers and Ulam was the first one concerning the stability of isometries and was generalized further by D.G. Bourgin [22] as follows.

Theorem 7. *Assume that E_1 is a Banach space and E_2 is a uniformly convex Banach space. If a mapping $T : E_1 \to E_2$ satisfy $T(0) = 0$ and the inequality*

$$\left| \|T(x) - T(y)\| - \|x - y\| \right| < \varepsilon, \qquad x, y \in E_1, \tag{1.4}$$

for some $\varepsilon > 0$, then the limit

$$I(x) = \lim_{n \to \infty} \frac{T(2^n x)}{2^n}$$

exists for every $x \in E_1$ and the transformation I is an isometry and satisfies

$$\|T(x) - I(x)\| \le 12\varepsilon, \qquad x \in E_1.$$

Subsequently, D.H. Hyers and S.M. Ulam [77] studied a stability problem for spaces of continuous mappings and obtained the next theorem.

Theorem 8. *Let S_1 and S_2 be compact metric spaces and $C(S_i)$ denote the space of real-valued continuous mappings on S_i equipped with the supremum norm $\|\cdot\|_\infty$. If a homeomorphism $T : C(S_1) \to C(S_2)$ satisfies the inequality*

$$\left| \|T(f) - T(g)\|_\infty - \|f - g\|_\infty \right| \le \varepsilon, \qquad f, g \in C(S_1), \tag{1.5}$$

for some $\varepsilon > 0$, then there exists an isometry $I : C(S_1) \to C(S_2)$ such that

$$\|T(f) - I(f)\|_\infty \le 21\varepsilon, \qquad f \in C(S_1).$$

This result was significantly generalized by D.G. Bourgin in [23] in the following way.

Theorem 9. *Let S_1 and S_2 be completely regular Hausdorff spaces and let $T : C(S_1) \to C(S_2)$ be a surjective mapping satisfying inequality (1.5) for some $\varepsilon > 0$. Then there exists a linear isometry $I : C(S_1) \to C(S_2)$ such that*

$$\|T(f) - I(f)\|_\infty \le 10\varepsilon, \qquad f \in C(S_1).$$

The study of stability problems for isometries on finite-dimensional Banach spaces was continued by R.D. Bourgin [25].

In 1978, P.M. Gruber [72] obtained an elegant result as follows.

Theorem 10. *Let E_1 and E_2 be real normed spaces and $T : E_1 \to E_2$ be a surjective mapping, which satisfies inequality (1.4) for some $\varepsilon > 0$. Furthermore, assume that $I : E_1 \to E_2$ is an isometry with $T(p) = I(p)$ for some $p \in E_1$. If*

$$\|T(x) - I(x)\| = o(\|x\|)$$

as $\|x\| \to \infty$ uniformly, then I is a surjective linear isometry and

$$\|T(x) - I(x)\| \le 5\varepsilon, \qquad x \in E_1.$$

If, in addition T, is continuous, then

$$\|T(x) - I(x)\| \leq 3\varepsilon, \qquad x \in E_1.$$

J. Gevirtz [71] established the following stability result for isometries between arbitrary Banach spaces.

Theorem 11. *Given real Banach spaces E_1 and E_2, let $T : E_1 \to E_2$ be a surjective mapping satisfying inequality (1.4) for some $\varepsilon > 0$. Then there exists a surjective isometry $I : E_1 \to E_2$ such that*

$$\|T(x) - I(x)\| \leq 5\varepsilon, \qquad x \in E_1. \tag{1.6}$$

Later, M. Omladič and P. Šemrl [114] showed that the bound 5ε in (1.6) can be replaced by 2ε.

Recently, G. Dolinar [57] proved a somewhat surprising stability result for isometries. Namely, he showed that, for each real $p > 1$ and every finite-dimensional real Banach spaces E_1 and E_2, every surjective $T : E_1 \to E_2$ satisfying the inequality

$$\Big| \|T(x) - T(y)\| - \|x - y\| \Big| \leq \varepsilon \|x - y\|^p, \qquad x, y \in E_1,$$

for some $\varepsilon > 0$, must be an isometry.

On the other hand, R.L. Swain [141] considered the stability of isometries on bounded metric spaces and proved the following result.

Theorem 12. *Let M be a subset of a compact metric space (E, d) and let $\varepsilon > 0$ be given. Then there exists a $\delta > 0$ such that if $T : M \to E$ satisfies the inequality*

$$|d(T(x), T(y)) - d(x, y)| < \delta, \qquad x, y \in M, \tag{1.7}$$

then there exists an isometry $I : M \to E$ with

$$d(T(x), I(x)) < \varepsilon, \qquad x \in M.$$

Finally, let us recall that the subsequent stability result for isometries on bounded subsets of $\mathbb{R}^n$ was obtained by J.W. Fickett [62].

Theorem 13. *Let n be a positive integer. For $t \geq 0$ we write*

$$K_0(t) = K_1(t) = t, \qquad K_2(t) = 3\sqrt{3}t, \qquad K_i(t) = 27t^{m(i)},$$

where $m(i) = 2^{1-i}$ for $i \geq 3$. Suppose that M is a bounded subset of $\mathbb{R}^n$, with diameter $d(M)$, and

$$3K_n(\varepsilon/d(M)) \leq 1$$

for some $\varepsilon > 0$. If a mapping $T : M \to \mathbb{R}^n$ satisfies inequality (1.7), then there exists an isometry $I : M \to \mathbb{R}^n$ such that

$$|T(x) - I(x)| \leq d(M)K_{n+1}(\varepsilon/d(M)), \qquad x \in M.$$

Recently, S.M. Jung and B. Kim [87] investigated the stability of isometries on restricted domains. For further information on the stability of isometries and related topics, we refer to [7, 18, 99, 128, 130, 139].

4. Other functional equations and inequalities in several variables

Now, we present examples of stability results that have been obtained for various functional equations and inequalities in several variables.

Let us start with a result of D.H. Hyers and S.M. Ulam [78] for convex functions.

Theorem 14. *Let $D \subset \mathbb{R}^n$ be an open convex set with non-empty interior and a function $f : D \to \mathbb{R}$ satisfy*

$$f(tx + (1 - t)y) \leq tf(x) + (1 - t)f(y) + \varepsilon, \qquad x, y \in D, 0 \leq t \leq 1.$$

Let B be a closed bounded convex subset of D. Then there exists a convex function $g : B \to \mathbb{R}$ such that

$$|f(x) - g(x)| \leq k_n\varepsilon, \qquad x \in B,$$

where

$$k_n = \frac{n^2 + 3n}{4n + 4}.$$

Next, let us recall a result of Z. Kominek [93] for the Jensen equation.

Theorem 15. *Let $D \subset \mathbb{R}^n$ be a bounded set with a nonempty interior and Y be a Banach space. Assume that there exists $x_0 \in D$ such that $D_0 := D - x_0$ satisfies the condition*

$$\frac{1}{2}D_0 \subset D_0.$$

Then, for each function $f : D \to Y$ such that

$$\sup_{x,y \in D} \left\| f\left(\frac{x + y}{2}\right) - \frac{f(x) + f(y)}{2} \right\| < \infty,$$

there exists $g : \mathbb{R}^n \to Y$ such that

$$g\left(\frac{x + y}{2}\right) = \frac{g(x) + g(y)}{2}, \qquad x, y \in \mathbb{R}^n,$$

and

$$\sup_{x \in D} \|f(x) - g(x)\| < \infty.$$

The first author treating Hyers-Ulam stability of the quadratic equation was F. Skof [138], who proved the following.

Theorem 16. *Let X be a normed vector space and Y a Banach space. If a function $f : X \to Y$ fulfills*

$$\|f(x + y) + f(x - y) - 2f(x) - 2f(y)\| \le \delta, \qquad x, y \in X,$$

for some $\delta > 0$, then for every $x \in X$ the limit

$$g(x) = \lim_{n \to \infty} \frac{f(2^n x)}{2^{2n}}$$

exists and g is the unique solution of the functional equation

$$g(x + y) + g(x - y) = 2g(x) + 2g(y), \qquad x, y \in X,$$

with

$$\|f(x) - g(x)\| \le \frac{\delta}{2}, \qquad x \in X.$$

Since then, the stability problem for the quadratic equation has been extensively investigated by a number of mathematicians in, e.g., [56, 80, 124, 127].

The Hyers-Ulam-Rassias stability of the pexiderized versions of the additive, Jensen, and quadratic equations, i.e., of the following three equations

$$f(x + y) = g(x) + h(y),$$
$$f\left(\frac{x + y}{2}\right) = \frac{g(x) + h(y)}{2},$$
$$f(x + y) + g(x - y) = h(x) + k(y),$$

has been studied in [79, 95, 96]. In particular the subsequent theorem has been proved in [96].

Theorem 17. *Let X be a normed space, Y be a Banach space, and let $f, g, h : X \to Y$ be mappings. Assume that $K \ge 0$ and $p \ne 1$ are real numbers with*

$$\|f(x + y) - g(x) - h(y)\| \le K(\|x\|^p + \|y\|^p), \qquad x, y \in X \setminus \{0\}.$$

Then there exists a unique additive mapping $T : X \to Y$ such that

$$\|f(x) - T(x) - f(0)\| \leq \frac{4K(3 + 3^p)}{2^p|3 - 3^p|}\|x\|^p, \qquad x \in X \setminus \{0\}.$$

5. Stability of functional equations in a single variable

For surveys of various stability results for functional equations in a single variable we refer to [3, 26, 32, 34, 38, 39, 44]. Here, we only present some examples of such outcomes.

For instance, J.A. Baker [20] discusses the Hyers-Ulam stability of the functional equations of the form

$$\varphi(x) = g(x)\varphi(f(x)) + h(x). \tag{1.8}$$

In particular, he has proved the following result.

Theorem 18. *Let S be a nonempty set and X be a real (or complex) Banach space. Consider $f : S \to S$, $h : S \to X$, $g : S \to \mathbb{R}$ (or $\mathbb{C}$) with*

$$|g(x)| \leq \lambda, \qquad x \in S,$$

for some $0 \leq \lambda < 1$. Suppose that $\varphi_s : S \to X$ satisfies

$$\|\varphi_s(x) - g(x)\varphi_s(f(x)) - h(x)\| \leq \delta \qquad x \in S,$$

where $\delta > 0$ is a constant. Then there exists a unique function $\varphi : S \to X$ that satisfies equation (1.8) and

$$\|\varphi_s(x) - \varphi(x)\| \leq \frac{\delta}{1 - \lambda}, \qquad x \in S.$$

A similar result has been obtained also when E is a Banach algebra.

To formulate the next result of a similar type, let us recall that a mapping $\gamma : \mathbb{R}_+ \to \mathbb{R}_+$ ($\mathbb{R}_+$ stands for the set of nonnegative reals) is called a *comparison function* if it is nondecreasing and

$$\lim_{n \to \infty} \gamma^n(t) = 0, \qquad t \in (0, \infty).$$

Now, we are in a position to present [67, Theorem 2.2].

Theorem 19. *Let S be a nonempty set, (X, d) be a complete metric space, $\xi : S \to S$, $F : S \times X \to X$. Assume also that*

$$d(F(t, u), F(t, v)) \leq \gamma(d(u, v)), \qquad t \in S, \ u, v \in X,$$

where $\gamma : \mathbb{R}_+ \to \mathbb{R}_+$ is a comparison function, and let $\varphi_s : S \to X$, $\delta > 0$ be such that

$$d(\varphi_s(t), F(t, \varphi_s(\xi(t)))) \leq \delta, \qquad t \in S. \tag{1.9}$$

Then there is a unique solution $\varphi : S \to X$ of the equation

$$\varphi(t) = F(t, \varphi(\xi(t))) \tag{1.10}$$

such that

$$\rho(\varphi_s, \varphi) := \sup_{t \in S} d(\varphi_s(t), \varphi(t)) < \infty.$$

Moreover,

$$\rho(\varphi_s, \varphi) - \gamma(\rho(\varphi_s, \varphi)) \leq \delta.$$

The following quite general stability results for difference equations have been obtained in [40, Theorems 1 and 2].

Theorem 20. *Let G be an abelian group, d be a complete and invariant metric in G, $a_n : G \to G$ be a continuous isomorphism for every $n \in \mathbb{N}_0$, $\{\varepsilon_n\}_{n \in \mathbb{N}_0} \subset (0, \infty)$, $\{\lambda_n\}_{n \in \mathbb{N}_0} \subset \mathbb{R}_+$, and $\{x_n\}_{n \in \mathbb{N}_0}$, $\{b_n\}_{n \in \mathbb{N}_0} \subset G$. Suppose that*

$$d(x_{n+1}, a_n(x_n) + b_n) \leq \varepsilon_n, \qquad n \in \mathbb{N}_0,$$

$$\liminf_{n \to \infty} \frac{\varepsilon_{n-1} \lambda_n}{\varepsilon_n} > 1 \tag{1.11}$$

and

$$d(a_n(x), a_n(y)) \geq \lambda_n d(x, y), \qquad x, y \in G, n \in \mathbb{N}_0. \tag{1.12}$$

Then there exists a unique sequence $\{y_n\}_{n \in \mathbb{N}_0} \subset G$ such that

$$y_{n+1} = a_n(y_n) + b_n, \qquad n \in \mathbb{N}_0 \tag{1.13}$$

and

$$d(x_n, y_n) \leq M \varepsilon_{n-1}, \qquad n \in \mathbb{N}, \tag{1.14}$$

with an $M \in \mathbb{R}_+$.

Theorem 21. *Let (X, d) be a metric space, $\{x_n\}_{n \in \mathbb{N}_0} \subset X$, $\{a_n\}_{n \in \mathbb{N}_0} \subset X^X$, $\{\varepsilon_n\}_{n \in \mathbb{N}_0} \subset (0, \infty)$, and*

$$d(x_{n+1}, a_n(x_n)) \leq \varepsilon_n, \qquad n \in \mathbb{N}_0.$$

Suppose that there exists $\{\lambda_n\}_{n\in\mathbb{N}_0} \subset \mathbb{R}_+$ with

$$\limsup_{n\to\infty} \frac{\varepsilon_{n-1}\lambda_n}{\varepsilon_n} < 1$$

and

$$d(a_n(x), a_n(y)) \le \lambda_n d(x, y), \qquad x, y \in X, n \in \mathbb{N}_0.$$

Then there exist a sequence $\{y_n\}_{n\in\mathbb{N}_0} \subset X$ and an $M > 0$ such that

$$y_{n+1} = a_n(y_n), \qquad n \in \mathbb{N}_0$$

and

$$d(x_n, y_n) \le M\varepsilon_{n-1} \qquad n \in \mathbb{N}. \tag{1.15}$$

Let S be a nonempty set, X be a Banach space over a field $\mathbb{K} \in \{\mathbb{R}, \mathbb{C}\}$, and the functions $F : S \to X$, $f : S \to S$ and $a_i : S \to \mathbb{K}$ for $i = 1, \dots, m$ be given. A natural generalization of a particular case of equation (1.8) is the linear functional equation of the order $m \in \mathbb{N}$ of the form

$$\varphi(f^m(x)) = \sum_{j=1}^{m} a_j(x)\varphi(f^{m-j}(x)) + F(x), \tag{1.16}$$

for the unknown functions $\varphi : S \to X$.

We refer the reader to [3, 26, 34, 38, 44] for surveys on stability results for that equation (with arbitrary m) and its generalizations (some recent outcomes can be found in [88, 89, 107]). Here, we only present two simplified results from [41, 42].

To this end we need the following hypothesis concerning the roots of the equations (for $x \in S$)

$$z^m - \sum_{j=1}^{m} a_j(x)z^{m-j} = 0. \tag{1.17}$$

($\mathcal{H}$) *Functions $r_1, \dots, r_m : S \to \mathbb{C}$ satisfy the conditions*

$$\prod_{i=1}^{m}(z - r_i(x)) = z^m - \sum_{j=1}^{m} a_j(x)z^{m-j}, \qquad x \in S, z \in \mathbb{C},$$

$$r_j(f(x)) = r_j(x), \qquad x \in S, \ j = 2, \dots, m.$$

Now we are in a position to present a result that follows from [42, Theorem 1].

Theorem 22. *Let $\varepsilon_0 : S \to \mathbb{R}_+$ and ($\mathcal{H}$) be valid. Assume that $0 \notin a_m(S)$ and $\varphi_s : S \to$*

X fulfils the inequality

$$\left\| \varphi_s(f^m(x)) - \sum_{j=1}^{m} a_j(x)\varphi_s(f^{m-j}(x)) - F(x) \right\| \le \varepsilon_0(x) \qquad x \in S.$$

Further, suppose that

$$\varepsilon_1(x) := \sum_{k=0}^{\infty} \frac{\varepsilon_0(f^k(x))}{\prod_{p=0}^{k} |r_1(f^p(x))|} < \infty, \qquad x \in S,$$

and

$$\varepsilon_j(x) := \sum_{k=0}^{\infty} \frac{\varepsilon_{j-1}(f^k(x))}{|r_j(x)|^{k+1}} < \infty, \qquad x \in S, j > 1.$$

Then equation (1.16) has a solution $\varphi : S \to X$ *with*

$$\|\varphi_s(x) - \varphi(x)\| \le \varepsilon_m(x), \qquad x \in S.$$

The next stability result, for a simplified form of (1.16) (with constant functions a_j), can be deduced from [41, Theorem 2].

Theorem 23. *Let* $\delta \in \mathbb{R}_+$, $d_0, ..., d_{m-1} \in \mathbb{K}$, $\varphi_s : S \to X$ *satisfy*

$$\left\| \varphi_s(f^m(x)) - \sum_{j=0}^{m-1} d_j\varphi_s(f^j(x)) - F(x) \right\| \le \delta, \qquad x \in S, \tag{1.18}$$

and $r_1, ..., r_m \in \mathbb{C}$ *denote the roots of the characteristic equation*

$$r^m - \sum_{j=0}^{m-1} d_j r^j = 0. \tag{1.19}$$

Assume that one of the following three conditions is valid:
1° $|r_j| > 1$ *for* $j = 1, ..., m$.
2° $|r_j| \in (1, \infty) \cup \{0\}$ *for* $j = 1, ..., m$ *and* f *is injective.*
3° $|r_j| \ne 1$ *for* $j = 1, ..., m$ *and* f *is bijective.*

Then there is a solution $\varphi : S \to X$ *of the equation*

$$\varphi(f^m(x)) = \sum_{j=0}^{m-1} d_j\varphi(f^j(x)) + F(x) \tag{1.20}$$

with

$$\|\varphi_s(x) - \varphi(x)\| \le \frac{\delta}{|1 - |r_1|| \cdot ... \cdot |1 - |r_m||}, \qquad x \in S. \tag{1.21}$$

Moreover, in the case where 1° or 3° holds, φ is the unique solution of (1.20) such that

$$\sup_{x \in S} \|\varphi_s(x) - \varphi(x)\| < \infty.$$

The issue of stability of functional equations in one variable has been investigated also for multi-valued functions; for suitable results and references we refer the reader to [39]. Here, we provide only one example of such results.

Let S be a nonempty set, (Y, d) be a metric space, and $n(Y)$ be the family of all nonempty subsets of Y. Write

$$\delta(A) := \sup \{d(x, y) : x, y \in A\}$$

for nonempty sets $A \subset Y$ and, given $F : S \to n(Y)$, denote by cl$F$ the multifunction defined by

$$(\mathrm{cl}F)(x) := \mathrm{cl}F(x), \qquad x \in S.$$

Remember that each $f : S \to Y$ with

$$f(x) \in F(x), \qquad x \in S,$$

is said to be a *selection* of the multifunction F.

The following result has been obtained in [115, Theorem 2] (see also [43]).

Theorem 24. *Let $F : S \to n(Y)$, $\Psi : Y \to Y$, $\xi : S \to S$, $\lambda \in (0, \infty)$,*

$$d(\Psi(x), \Psi(y)) \le \lambda d(x, y), \qquad x, y \in Y$$

and

$$\lim_{n \to \infty} \lambda^n \delta(F(\xi^n(x))) = 0, \qquad x \in S.$$

1) *If Y is complete and*

$$\Psi(F(\xi(x))) \subset F(x), \qquad x \in S,$$

then, for each $x \in S$, the limit

$$\lim_{n \to \infty} \mathrm{cl}(\Psi^n \circ F \circ \xi^n)(x) =: f(x)$$

exists (with respect to the Hausdorff distance in $n(Y)$) and f is a unique selection of the multifunction cl$F$ such that

$$\Psi \circ f \circ \xi = f.$$

2) *If*

$$F(x) \subset \Psi(F(\xi(x))), \qquad x \in S,$$

then F is a single-valued function and

$$\Psi \circ F \circ \xi = F.$$

6. Iterative stability

D. Brydak [27] (cf. [49, Definition 2]) introduced the notion of stability, which later has been called *iterative stability*. Namely, let $I = (0, d]$ for a $d > 0$, and $\xi : I \to I$, $a, h : I \to \mathbb{R}$ be given functions. For the linear equation

$$\varphi(\xi(x)) = a(x)\varphi(x) + h(x), \qquad x \in I, \qquad (1.22)$$

(for functions $\varphi : I \to \mathbb{R}$) it means that there exists a real constant $\kappa > 0$ such that, for every $\varepsilon > 0$ and every continuous function $\psi : I \to \mathbb{R}$ satisfying the condition

$$\left| \psi(\xi^n(x)) - G_n(x)\psi(x) - G_n(x) \sum_{i=0}^{n-1} \frac{h(\xi^i(x))}{G_{i+1}(x)} \right| \leq \varepsilon, \qquad x \in I, n \in \mathbb{N}, \qquad (1.23)$$

there exists a continuous solution φ of equation (1.22) such that

$$|\psi(x) - \varphi(x)| \leq \kappa\varepsilon, \qquad x \in I, \qquad (1.24)$$

where

$$G_n(x) := \prod_{i=0}^{n-1} a(\xi^i(x)), \qquad x \in I. \qquad (1.25)$$

D. Brydak [27] proved that if equation (1.22) has a continuous solution, the limit

$$G(x) := \lim_{n \to \infty} G_n(x)$$

exists for each $x \in I$, G is continuous in I,

$$G(x) \neq 0, \qquad x \in I,$$

and

$$\inf_{x \in I} |G(x)| > 0,$$

then equation (1.22) is iteratively stable.

For a further detailed discussion and references concerning iterative stability and similar issues we refer the reader to survey papers [3, 34].

7. Differential and integral equations

Let n be a positive integer, I be a nondegenerate interval of $\mathbb{R}$, and $F : \mathbb{R}^{n+1} \times I \to \mathbb{R}$. Consider, under suitable assumptions on F, stability of the differential equation of the *n*-th order

$$F(y^{(n)}(x), y^{(n-1)}(x), \ldots, y'(x), y(x), x) = 0. \tag{1.26}$$

Namely, for an arbitrary $\varepsilon > 0$, consider the differential inequality

$$|F(y^{(n)}(x), y^{(n-1)}(x), \ldots, y'(x), y(x), x)| \leq \varepsilon, \qquad x \in I, \tag{1.27}$$

for the n times continuously differentiable functions $y : I \to \mathbb{R}$. If for each function $y : I \to \mathbb{R}$ satisfying (1.27), there exists a solution $y_0 : I \to \mathbb{R}$ of (1.26) such that

$$|y(x) - y_0(x)| \leq K(\varepsilon), \qquad x \in I, \tag{1.28}$$

where $K(\varepsilon)$ depends on ε only and

$$\lim_{\varepsilon \to 0} K(\varepsilon) = 0,$$

then we say that the differential equation (1.26) is Hyers-Ulam stable (or has the Hyers-Ulam stability); if the domain I is not the whole space $\mathbb{R}$, then we say that it has the local Hyers-Ulam stability. When the above statement also holds with ε and $K(\varepsilon)$ replaced with some appropriate functions $\varphi, \Phi : I \to \mathbb{R}_+$ (nonnegative reals), respectively, then we say that the differential equation (1.26) has the generalized Hyers-Ulam stability (or the Hyers-Ulam-Rassias stability).

It seems that the Hyers-Ulam stability of differential equations was investigated first by M. Obłoza [113] in the following way.

Theorem 25. *Given real constants a and b, let $g, r : (a, b) \to \mathbb{R}$ be continuous functions with*

$$\int_a^b |g(x)|dx < \infty.$$

Assume that $\varepsilon > 0$ is an arbitrary real number. Assume that a differentiable function $y : (a, b) \to \mathbb{R}$ satisfies the inequality

$$|y'(x) + g(x)y(x) - r(x)| \leq \varepsilon, \qquad x \in (a, b),$$

and a function $y_0 : (a, b) \to \mathbb{R}$ is such that

$$y_0'(x) + g(x)y_0(x) = r(x), \qquad x \in (a, b),$$

and $y(\tau) = y_0(\tau)$ for some $\tau \in (a, b)$. Then there exists a constant $\delta > 0$ with

$$|y(x) - y_0(x)| \leq \delta, \qquad x \in (a, b).$$

Later C. Alsina and R. Ger [9] proved the subsequent result.

Theorem 26. *Let $a, b, \varepsilon \in \mathbb{R}$, $a < b$, and $\varepsilon > 0$. For every differentiable function $y :$*

$(a, b) \rightarrow \mathbb{R}$ *satisfying the differential inequality*

$$|y'(x) - y(x)| \leq \varepsilon, \qquad x \in (a, b),$$

there exists a differentiable function $y_0 : (a, b) \rightarrow \mathbb{R}$ *such that*

$$y_0'(x) = y_0(x), \qquad x \in (a, b),$$

$$|y(x) - y_0(x)| \leq 3\varepsilon, \qquad x \in (a, b).$$

The result of Alsina and Ger was extended by T. Miura, S. Miyajima, and S.E. Takahasi [104, 105, 142] and by S.E. Takahasi, H. Takagi, T. Miura, and S. Miyajima [143] to the first-order linear differential equations and linear differential equations of higher order with constant coefficients. In particular, the following result has been proved in [105].

Theorem 27. *Let X be a non-zero complex Banach space with a norm* $\| \cdot \|$, $D = d/dt$ *be the differential operator, and* $P(z)$ *a polynomial of degree* $n \in \mathbb{N}$ *with complex coefficients. Then the following conditions are equivalent.*

(i) $P(z) = 0$ *has no purely imaginary solutions.*
(ii) $P(D) : C^n(\mathbb{R}, X) \rightarrow C(\mathbb{R}, X)$ *has the Hyers-Ulam stability.*
(iii) The equation $P(D)f = 0$ *has the Hyers-Ulam stability.*

Moreover, if $P(D)$ *has the Hyers-Ulam stability, then for each* $\varepsilon \geq 0$, $g \in C(\mathbb{R}, X)$ *and* $f \in C^n(\mathbb{R}, X)$ *with*

$$\|P(D)f - g\|_\infty \leq \varepsilon,$$

the element $f_0 \in C^n(\mathbb{R}, X)$, *satisfying the conditions*

$$P(D)f_0 = g, \qquad \|f - f_0\|_\infty < \infty,$$

is uniquely determined, where

$$\|f\|_\infty := \sup_{t \in \mathbb{R}} \|f(t)\|, \qquad f \in C(\mathbb{R}, X).$$

Extensions of the outcomes of Takahasi, Takagi, and Miura have been obtained by S.M. Jung [82, 83, 84, 91]. Let us recall here a result in [83].

Theorem 28. *Let X be a complex Banach space and let* $I = (a, b)$ *be an open interval, where* $a, b \in \mathbb{R} \cup \{\pm\infty\}$ *are arbitrarily given with* $a < b$. *Assume that* $g : I \rightarrow \mathbb{C}$ *and* $h : I \rightarrow X$ *are continuous functions such that* $g(t)$ *and the function*

$$t \rightarrow h(t) \exp\left(\int_a^t g(u)du\right)$$

are integrable on (a, c) for each $c \in I$. Moreover, suppose $\varphi : I \to [0, \infty)$ is a function such that the function

$$t \to \varphi(t) \exp\left(\Re\left(\int_a^t g(u)du\right)\right)$$

is integrable on I. If a continuously differentiable function $y : I \to X$ satisfies the differential inequality

$$\|y'(t) + g(t)y(t) + h(t)\| \leq \varphi(t)$$

for all $t \in I$, then there exists a unique solution $y_0 : I \to X$ of the differential equation

$$y'(t) + g(t)y(t) + h(t) = 0$$

such that

$$\|y(t) - y_0(t)\| \leq \exp\left(-\Re\left(\int_a^t g(u)du\right)\right) \int_t^b \varphi(v) \exp\left(\Re\left(\int_a^v g(u)du\right)\right) dv$$

for every $t \in I$.

I.A. Rus [134, 135] obtained some results on the stability of differential equations using Gronwall lemma and the technique of weakly Picard operators. Recently, G. Wang, M. Zhou, and L. Sun [145] and Y. Li and Y. Shen [98] proved the Hyers-Ulam stability of the linear differential equation of the first order and the linear differential equation of the second order with constant coefficients by using the method of integral factor. For some related outcomes we refer to [35, 147, 148].

Some extensions of the results given in [84, 98, 105] were obtained by D.S. Cîmpean and D. Popa, and by D. Popa and I. Raşa X for the linear differential equations of the n-th order (see [54, 120, 121]). It seems that the first paper on Hyers-Ulam stability of partial differential equations was written by A. Prástaro and Th.M. Rassias [122]; for recent results on this subject we refer the reader to [1, 55, 100, 101, 133, 137].

The issue of Hyers-Ulam-Rassias stability and Hyers-Ulam stability have also been investigated for integral equations, and for suitable information we refer to [6, 35, 46, 47, 64, 65, 69, 85, 103, 106, 136]. Here we present only few such outcomes. To this end, let us recall that, for a given continuous function f and a fixed real number c, the integral equation

$$y(x) = \int_c^x f(\tau, y(\tau))d\tau \tag{1.29}$$

is called a Volterra integral equation of the second kind. If for each function $y(x)$

satisfying

$$\left| y(x) - \int_c^x f(\tau, y(\tau)) d\tau \right| \leq \psi(x) \tag{1.30}$$

where $\psi(x) \geq 0$ for all x, there exists a solution $y_0(x)$ of the Volterra integral equation (1.29) and a constant $C > 0$ with

$$|y(x) - y_0(x)| \leq C\psi(x) \tag{1.31}$$

for all x, where C is independent of $y(x)$ and $y_0(x)$; then we say that the integral equation (1.29) has the Hyers-Ulam-Rassias stability. If $\psi(x)$ is a constant function in the above inequalities, we say that the integral equation (1.29) has the Hyers-Ulam stability.

For instance, S.M. Jung [85] has proved the following result on the Hyers-Ulam-Rassias stability and the Hyers-Ulam stability of integral equation (1.29).

Theorem 29. *Let K and L be positive constants with $0 < KL < 1$ and let $I = [a, b]$ be given for fixed real numbers a, b with $a < b$. Assume that $f : I \times \mathbb{C} \to \mathbb{C}$ is a continuous function which satisfies a Lipschitz condition,*

$$|f(x, y) - f(x, z)| \leq L|y - z|, \tag{1.32}$$

for any $x \in I$ and all $y, z \in \mathbb{C}$. If a continuous function $y : I \to \mathbb{C}$ satisfies

$$\left| y(x) - \int_c^x f(\tau, y(\tau)) d\tau \right| \leq \psi(x)$$

for all x and for some $c \in I$, where $\psi : I \to (0, \infty)$ is a continuous function with

$$\left| \int_c^x \psi(\tau) d\tau \right| \leq K\psi(x)$$

for each $x \in I$, then there exists a unique continuous function $y_0 : I \to \mathbb{C}$ such that

$$y_0(x) = \int_c^x f(\tau, y_0(\tau)) d\tau,$$

$$|y(x) - y_0(x)| \leq \frac{1}{1 - KL} \psi(x)$$

for all $x \in I$.

Theorem 30. *Given $a \in \mathbb{R}$ and $r > 0$, let $I(a; r)$ denote a closed interval $\{x \in \mathbb{R} | a - r \leq x \leq a + r\}$ and let $f : I(a; r) \times \mathbb{C} \to \mathbb{C}$ be a continuous function that satisfies a Lipschitz condition (1.32) for all $x \in I(a; r)$ and $y, z \in \mathbb{C}$, where L is a constant with*

$0 < Lr < 1$. *If a continuous function* $y : I(a; r) \to \mathbb{C}$ *satisfies*

$$\left| y(x) - b - \int_c^x f(\tau, y(\tau)) d\tau \right| \leq \theta$$

for all $x \in I(a; r)$ *and for some* $\theta \geq 0$, *where b is a complex number, then there exists a unique continuous function* $y_0 : I(a; r) \to \mathbb{C}$ *such that*

$$y_0(x) = b + \int_c^x f(\tau, y_0(\tau)) d\tau,$$

$$|y(x) - y_0(x)| \leq \frac{\theta}{1 - Lr}$$

for all $x \in I(a; r)$.

In 2009 L.P. Castro and A. Ramos [46] investigated Hyers-Ulam stability for a generalized Volterra integral equation of the form

$$y(x) = \int_a^x f(x, \tau, y(\tau)) d\tau \tag{1.33}$$

on a finite and on an infinite intervals.

Theorem 31. *Let K and L be positive constants with* $0 < KL < 1$ *and assume that* $f : [a, b] \times [a, b] \times \mathbb{C} \to \mathbb{C}$ *is a continuous function that additionally satisfies the Lipschitz condition*

$$|f(x, \tau, y) - f(x, \tau, z)| \leq L|y - z| \tag{1.34}$$

for any $x, \tau \in [a, b]$ *and all* $y, z \in \mathbb{C}$. *If a continuous function* $y : [a, b] \to \mathbb{C}$ *satisfies*

$$\left| y(x) - \int_a^x f(x, \tau, y(\tau)) d\tau \right| \leq \psi(x) \tag{1.35}$$

for all $x \in [a, b]$, *and where* $\psi : [a, b] \to (0, \infty)$ *is a continuous function with*

$$\left| \int_a^x \psi(\tau) d\tau \right| \leq K\psi(x) \tag{1.36}$$

for each $x \in [a, b]$, *then there exists a unique continuous function* $y_0 : [a, b] \to \mathbb{C}$ *such that*

$$y_0(x) = \int_a^x f(x, \tau, y_0(\tau)) d\tau, \tag{1.37}$$

$$|y(x) - y_0(x)| \leq \frac{1}{1 - KL}\psi(x) \tag{1.38}$$

for all $x \in [a, b]$.

Theorem 32. *Let K and L be positive constants with $0 < KL < 1$ and assume $f : \mathbb{R} \times \mathbb{R} \times \mathbb{C} \to \mathbb{C}$ is a continuous function that additionally satisfies the Lipschitz condition (1.34), for any $x, \tau \in \mathbb{R}$ and all $y, z \in \mathbb{C}$. If a continuous function $y : \mathbb{R} \to \mathbb{C}$ satisfies (1.35), for all $x \in \mathbb{R}$ and for some $a \in \mathbb{R}$, where $\psi : \mathbb{R} \to (0, \infty)$ is a continuous function satisfying (1.36), for each $x \in \mathbb{R}$, then there exists a unique continuous function $y_0 : \mathbb{R} \to \mathbb{C}$ which satisfies (1.37) and (1.38) for all $x \in \mathbb{R}$.*

Let us mention that M. Gachpazan and O. Baghani [64] proved a result on the Hyers-Ulam stability of the following nonhomogeneous nonlinear Volterra integral equation:

$$y(x) = f(x) + \varphi\left(\int_a^x F(x, \tau, y(\tau))d\tau \right),$$

where $x \in [a, b]$, with $-\infty < a < b < \infty$.

8. Superstability and hyperstability

We say that a functional equation is superstable when an unbounded approximate solution to the equation must be a true solution to it. The first superstability results were obtained by D.G. Bourgin [23] in connection with the notion of approximate isometries. Next, J.A. Baker, J. Lawrence, F. Zorzitto [21], and J.A. Baker [19] proved the superstability of the exponential equation

$$f(x + y) = f(x)f(y).$$

Here, we recall the result of J.A. Baker [19] (see also [11, 90] for more information on some recent related investigations).

Theorem 33. *Let $\delta > 0$, S be a semigroup, and f be a complex-valued function defined on S such that*

$$|f(xy) - f(x)f(y)| \le \delta, \qquad x, y \in S.$$

Then either

$$|f(x)| \le (1 + \sqrt{1 + 4\delta})/2, \qquad x \in S,$$

or

$$f(xy) = f(x)f(y), \qquad x, y \in S.$$

Using that result and a theorem of Pl. Kannappan [92], J.A. Baker [19] proved further the following superstability outcome for the cosine equation.

Theorem 34. *Let $\delta > 0$, G be an abelian group, and f be a complex-valued function defined on G such that*

$$|f(x+y) + f(x-y) - 2f(x)f(y)| \le \delta, \qquad x, y \in G.$$

Then either

$$|f(x)| \le \frac{1 + \sqrt{1 + 2\delta}}{2}, \qquad x \in G,$$

or

$$f(x+y) + f(x-y) = 2f(x)f(y), \qquad x, y \in G.$$

Since then, numerous similar results have been published and we refer to the survey [33] for more information and references concerning the notions of superstability and hyperstability (which quite often are confused).

Probably the first hyperstability result was published in [23] and concerned ring homomorphisms. However, the term "hyperstability" was used for the first time in [102].

As mentioned before, the following extension of Theorem 3 is valid.

Theorem 35. *Let E_1 and E_2 be normed spaces, E_2 complete, and $K \ge 0$ and $p \ne 1$ fixed real numbers. If $f : E_1 \to E_2$ is a mapping satisfying*

$$\|f(x+y) - f(x) - f(y)\| \le K(\|x\|^p + \|y\|^p), \qquad x, y \in E_1 \setminus \{0\}, \tag{1.39}$$

then there exists a unique function $g : E_1 \to E_2$ such that

$$g(x+y) = g(x) + g(y), \qquad x, y \in E_1,$$

$$\|f(x) - g(x)\| \le \frac{K\|x\|^p}{|2^{p-1} - 1|}, \qquad x \in E_1 \setminus \{0\}.$$

It has a very nice simple form, but it has been improved in [94] to show that, in the case $p < 0$, each $f : E_1 \to E_2$ satisfying (1.39) must actually be additive (and the completeness of E_2 is not necessary in such a situation). Below, we present a slightly more general theorem from [28].

Theorem 36. *Let E_1 and E_2 be normed spaces, $X \subset E_1 \setminus \{0\}$ nonempty, $K \ge 0$, and $p < 0$. Assume also that*

$$-X = X,$$

where $-X := \{-x : x \in X\}$, and there exists a positive integer m_0 with

$$-x, nx \in X, \qquad x \in X, n \in \mathbb{N}, n \geq m_0.$$

Then every operator $g : E_1 \to E_2$ such that

$$\|g(x + y) - g(x) - g(y)\| \leq K(\|x\|^p + \|y\|^p), \qquad x, y \in X, x + y \in X,$$

is additive on X; that is,

$$g(x + y) = g(x) + g(y), \qquad x, y \in X, x + y \in X.$$

Theorem 36 presents a simple particular observation on Φ−hyperstability for the Cauchy equation on a restricted domain, with

$$\Phi(x, y) = K(\|x\|^p + \|y\|^p), \qquad x, y \in E_1,$$

for a fixed real $p < 0$ and $K > 0$ (for some further similar outcomes we refer to [29, 30]). Generally, given $\Phi : E_1^2 \to [0, \infty)$ (under the assumptions of Theorem 36), we say that the conditional functional equation

$$g(x + y) = g(x) + g(y), \qquad x, y \in X, x + y \in X, \tag{1.40}$$

is Φ-hyperstable in the class of functions $f : X \to E_2$ provided each $g : X \to E_2$ satisfying the inequality

$$\|g(x + y) - g(x) - g(y)\| \leq \Phi(x, y), \qquad x, y \in X, x + y \in X,$$

must be additive on X, that is (1.40) holds.

The hyperstability of some other functional equations has been discussed successively. For instance, M. Piszczek in [116] has studied the hyperstability of the linear equation

$$g(ax + by) = Ag(x) + Bg(y).$$

Moreover, A. Bahyrycz and J. Olko [15] discussed the hyperstability of the general linear equation

$$\sum_{i=1}^{m} A_i g\left(\sum_{j=1}^{n} a_{ij} x_j \right) + A = 0.$$

For further examples of quite recent analogous hyperstability investigations we refer to [4, 5, 8, 12, 16, 17, 31, 58, 59, 60, 117, 118, 149].

9. Composite type equations

The investigation of stability of the composite functional equations has been motivated by a question R. Ger asked in 2000 (at the 38th International Symposium on Functional Equations), concerning in particular the Hyers-Ulam stability of the Gołąb-Schinzel equation

$$f(x + f(x)y) = f(x)f(y). \tag{1.41}$$

The first result of this type has been proved by J. Chudziak [50], who showed that if $f : \mathbb{R} \to \mathbb{R}$ is a continuous function satisfying

$$|f(x + f(x)y) - f(x)f(y))| \le \varepsilon, \qquad x, y \in \mathbb{R}, \tag{1.42}$$

with a positive real number ε, then either f is bounded or it is a solution of (1.41).

Next, J. Chudziak and J. Tabor [53] generalized this result, by proving that if $\mathbb{K}$ is a subfield of $\mathbb{C}$, X is a vector space over $\mathbb{K}$ and $f : X \to \mathbb{K}$ is a function with

$$|f(x + f(x)y) - f(x)f(y))| \le \varepsilon, \qquad x, y \in X, \tag{1.43}$$

and such that the limit

$$\lim_{t \to 0} f(tx) \tag{1.44}$$

exists (not necessarily finite) for every $x \in X \setminus f^{-1}(0)$, then either f is bounded or it is a solution of (1.41) on X.

Later on, in [51] and [52], analogous results have been proved for the equation

$$f(x + f(x)^n y) = \lambda f(x)f(y) \tag{1.45}$$

where n is a positive integer and λ is a nonzero complex number.

A survey of the stability results for the functional equations of Gołąb-Schinzel type can be found in [2]. Stability of few other similar equations have been investigated in [37, 48, 61, 112]. For a survey on stability outcomes for the translation equation

$$F(t, F(s, x)) = F(s + t, x)$$

we refer to [123].

10. Nonstability

In connection with the stability result in [125], depicted by Theorem 3, and an easy observation that its proof actually works for $p < 0$, Th.M. Rassias asked a natural question about an analogous result for $p \ge 1$. This problem was raised by him during the 27th International Symposium on Functional Equations (cf. [126]). Very soon Z. Gajda [66] gave an affirmative answer to the question for $p > 1$ (using the direct method) and provided a simple counterexample showing that this is not the case for

$p = 1$. Namely, he showed that, for each $K > 0$, there exists a function $f : \mathbb{R} \to \mathbb{R}$ such that

$$|f(x + y) - f(x) - f(y)| \le K(|x| + |y|), \qquad x, y \in \mathbb{R}, \qquad (1.46)$$

and there does not exist any constant $\widetilde{K} \in [0, \infty)$ and any additive function $g : \mathbb{R} \to \mathbb{R}$ with

$$|f(x) - g(x)| \le \widetilde{K}|x|, \qquad x \in \mathbb{R}. \qquad (1.47)$$

The example is depicted below.

Example 1. Fix $\mu > 0$ and define a function $\phi : \mathbb{R} \to \mathbb{R}$ by

$$\phi(x) := \begin{cases} \mu, & x \in [1, \infty), \\ \mu x, & x \in (-1, 1), \\ -\mu, & x \in (-\infty, -1]. \end{cases} \qquad (1.48)$$

Clearly, ϕ is continuous and $|\phi(x)| \le \mu$ for all $x \in \mathbb{R}$. So, we may define a continuous function $f : \mathbb{R} \to \mathbb{R}$ by

$$f(x) := \sum_{n=0}^{\infty} \frac{\phi(2^n x)}{2^n}, \qquad x \in \mathbb{R},$$

Moreover,

$$|f(x)| \le \sum_{n=0}^{\infty} \frac{\mu}{2^n} = 2\mu, \qquad x \in \mathbb{R}.$$

It can be shown (see [66]) that (1.46) holds with $K := 6\mu$ and for every $\widetilde{K} \in [0, \infty)$ and every additive function $g : \mathbb{R} \to \mathbb{R}$ condition (1.47) does not hold.

After that Th.M. Rassias and P. Šemrl [131] have noticed that the unbounded continuous function $f : \mathbb{R} \to \mathbb{R}$, defined by

$$f(x) = \begin{cases} x \log_2(x + 1), & x \ge 0, \\ x \log_2 |x - 1|, & x < 0, \end{cases}$$

satisfies the inequality

$$|f(x + y) - f(x) - f(y)| \le |x| + |y|, \qquad x \in \mathbb{R},$$

and $f(x)/x \to \infty$ as $x \to \infty$, whence no continuous additive function $g : \mathbb{R} \to \mathbb{R}$ can satisfy the condition

$$\sup_{x \in \mathbb{R}_0} \frac{|f(x) - g(x)|}{|x|} < \infty,$$

where $\mathbb{R}_0 := \mathbb{R} \setminus \{0\}$.

Further issues connected with the notion of nonstability are discussed in the last chapter of this book.

REFERENCES

1. M.R. Abdollahpour, R. Aghayari, M.Th. Rassias, Hyers-Ulam stability of associated Laguerre differential equations in a subclass of analytic functions, J. Math. Anal. Appl. 437 (2016) 605–612.
2. R.P. Agarwal, J. Brzdęk, J. Chudziak, Stability problem for the composite type functional equations, Expo. Math., to appear.
3. R.P. Agarwal, B. Xu, W. Zhang, Stability of functional equations in single variable, J. Math. Anal. Appl. 288 (2003) 852–869.
4. L. Aiemsomboon, W. Sintunavarat, On generalized hyperstability of a general linear equation, Acta Math. Hungar. 149 (2016) 413–422.
5. L. Aiemsomboon, W. Sintunavarat, Two new generalised hyperstability results for the Drygas functional equation, Bull. Aust. Math. Soc. 95 (2017) 269–280.
6. M. Akkouchi, Hyers-Ulam-Rassias stability of nonlinear Volterra integral equations via a fixed point approach, Acta Univ. Apulensis Math. Inform. 26 (2011) 257–266.
7. P. Alestalo, Isometric approximation in bounded sets and its applications, In: Developments in Functional Equations and Related Topics (J. Brzdęk, K. Ciepliński, Th. M. Rassias, eds.), Springer Optimization and Its Applications, to appear.
8. M. Almahalebi, On the hyperstability of σ-Drygas functional equation on semigroups, Aequationes Math. 90 (2016) 849–857.
9. C. Alsina, R. Ger, On some inequalities and stability results related to the exponential function, J. Inequal. Appl. 2 (1998) 373–380.
10. T. Aoki, On the stability of the linear transformation in Banach spaces, J. Math. Soc. Japan 2 (1950) 64–66.
11. R. Badora, J. Chmieliński, Decomposition of mappings approximately inner product preserving, Nonlinear Anal. TMA 62 (2005) 1015–1023.
12. A. Bahyrycz, Hyperstability of some functional equation on restricted domain: direct and fixed point methods, Bull. Iran. Math. Soc. 42 (2016) 959–974.
13. A. Bahyrycz, J. Brzdęk, Remarks on stability of the equation of homomorphism for square symmetric groupoids, In: (Th. M Rassias, ed.) Handbook of Functional Equations: Stability Theory, pp. 37–57, Springer Optimization and Its Applications, Springer, 2014.
14. A. Bahyrycz, K. Ciepliński, On an equation characterizing multi-Jensen-quadratic mappings and its Hyers-Ulam stability via a fixed point method, J. Fixed Point Theory Appl. 18 (2016) 737–751.
15. A. Bahyrycz, J. Olko, Hyperstability of general linear functional equation. Aequationes Math. 90 (2016) 527–540.
16. A. Bahyrycz, Zs. Páles, M. Piszczek, Asymptotic stability of the Cauchy and Jensen functional equations, Acta Math. Hungar. 150 (2016) 131–141.
17. A. Bahyrycz, M. Piszczek, Hyperstability of the Jensen functional equation, Acta Math. Hungar. 142 (2014) 353–365.
18. J.A. Baker, Isometries in normed spaces, Amer. Math. Monthly 78 (1971) 655–658.
19. J.A. Baker, The stability of the cosine equation, Proc. Amer. Math. Soc. 80 (1980) 411–416.
20. J.A. Baker, The stability of certain functional equations, Proc. Amer. Math. Soc. 112 (1991) 729–732.
21. J.A. Baker, J. Lawrence, F. Zorzitto, The stability of the equation $f(x + y) = f(x)f(y)$, Proc. Amer. Math. Soc. 74 (1979) 242–246.
22. D.G. Bourgin, Approximate isometries, Bull. Amer. Math. Soc. 52 (1946) 704–714.
23. D.G. Bourgin, Approximately isometric and multiplicative transformations on continuous function rings, Duke Math. J. 16 (1949) 385–397.
24. D.G. Bourgin, Classes of transformations and bordering transformations, Bull. Amer. Math. Soc. 57

(1951) 223–237.

25. R.D. Bourgin, Approximate isometries on finite dimensional Banach spaces, Trans. Amer. Math. Soc. 207 (1975) 309–328.

26. N. Brillouët-Belluot, J. Brzdęk, K. Ciepliński, On some recent developments in Ulam's type stability, Abstr. Appl. Anal. 2012 (2012), Art. ID 716936, 41 pp.

27. D. Brydak, On the stability of the functional equation $\varphi[f(x)] = g(x)\varphi(x) + F(x)$, Proc. Amer. Math. Soc. 26 (1970) 455–460.

28. J. Brzdęk, Hyperstability of the Cauchy equation on restricted domains, Acta Math. Hungar. 141 (2013) 58–67.

29. J. Brzdęk, Remarks on hyperstability of the Cauchy functional equation, Aequationes Math. 86 (2013) 255–267.

30. J. Brzdęk, A hyperstability result for the Cauchy equation, Bull. Aust. Math. Soc. 89 (2014) 33–40.

31. J. Brzdęk, Remarks on stability of some inhomogeneous functional equations, Aequationes Math. 89 (2015) 83–96.

32. J. Brzdęk, L. Cădariu, K. Ciepliński, Fixed point theory and the Ulam stability, J. Funct. Space. 2014 (2014), Art. ID 829419, 16 pp.

33. J. Brzdęk, K. Ciepliński, Hyperstability and superstability, Abstr. Appl. Anal. 2013 (2013), Art. ID 401756, 13 pp.

34. J. Brzdęk, K. Ciepliński, Z. Leśniak, On Ulam's type stability of the linear equation and related issues, Discrete Dyn. Nat. Soc. 2014 (2014), Art. ID 536791, 14 pp.

35. J. Brzdęk, N. Eghbali, On approximate solutions of some delayed fractional differential equations, Appl. Math. Lett. 54 (2016) 31–35.

36. J. Brzdęk, W. Fechner, M.S. Moslehian, J. Sikorska, Recent developments of the conditional stability of the homomorphism equation, Banach J. Math. Anal. 9 (2015) 278–326.

37. J. Brzdęk, A. Najdecki, B. Xu, Two general theorems on superstability of functional equations. Aequationes Math. 89 (2015) 771–783.

38. J. Brzdęk, M. Piszczek, On stability of the linear and polynomial functional equations in single variable, In: (Th. M Rassias, ed.) Handbook of Functional Equations: Stability Theory, pp. 59–81, Springer Optimization and Its Applications, Springer, 2014.

39. J. Brzdęk, M. Piszczek, Selections of set-valued maps satisfying some inclusions and the Hyers-Ulam stability, In: (Th. M Rassias, ed.) Handbook of Functional Equations: Stability Theory, pp. 83–100, Springer Optimization and Its Applications, Springer, 2014.

40. J. Brzdęk, D. Popa, B. Xu, The Hyers-Ulam stability of nonlinear recurrences, J. Math. Anal. Appl. 335 (2007) 443–449.

41. J. Brzdęk, D. Popa, B. Xu, Hyers-Ulam stability for linear equations of higher orders, Acta Math. Hungar. 120 (2008) 1–8.

42. J. Brzdęk, D. Popa, B. Xu, On approximate solutions of the linear functional equation of higher order, J. Math. Anal. Appl. 373 (2011) 680–689.

43. J. Brzdęk, D. Popa, B. Xu, Selections of set-valued maps satisfying a linear inclusion in a single variable, Nonlinear Anal. TMA 74 (2011) 324–330.

44. J. Brzdęk, D. Popa, B. Xu, Remarks on stability of the linear functional equation in single variable. In: (P.M. Pardalos, P.G. Georgiev, H.M. Srivastava, eds.) Nonlinear Analysis: Stability, Approximation, and Inequalities, pp. 91–119, Springer Optimization and Its Applications, vol. 68, Springer, New York, 2012.

45. J. Brzdęk, J. Sikorska, A conditional exponential functional equation and its stability, Nonlinear Anal. TMA 72 (2010) 2923–2934.

46. L.P. Castro, A. Ramos, Hyers-Ulam-Rassias stability for a class of nonlinear Volterra integral equations, Banach J. Math. Anal. 3 (2009) 36–43.

47. L.P. Castro, A. Ramos, Hyers-Ulam and Hyers-Ulam-Rassias stability of Volterra integral equations with delay, In: Integral methods in science and engineering, Vol. 1, pp. 85-94, Birkhäuser, Boston 2010.

48. A. Charifi, B. Bouikhalene, S. Kabbaj, J.M. Rassias, On the stability of a Pexiderized Gołąb-Schinzel equation, Comput. Math. Appl. 59 (2010) 3193–3202.

49. B. Choczewski, E. Turdza, R. Węgrzyk, On the stability of a linear functional equation, Wyż. Szkoła Ped. Krakow. Rocznik Nauk.-Dydakt. Prace Mat. 9 (1979) 15–21.

50. J. Chudziak, On a functional inequality related to the stability problem for the Gołąb-Schinzel equation, Publ. Math. Debrecen 67 (2005) 199–208.

51. J. Chudziak, Stability of the generalized Gołąb-Schinzel equation, Acta Math. Hungar. 113 (2006) 133–144.

52. J. Chudziak, Approximate solutions of the generalized Gołąb-Schinzel equation, J. Inequal. Appl. 2006, Art. ID 89402, 8 pp.

53. J. Chudziak, J. Tabor, On the stability of the Gołąb-Schinzel functional equation, J. Math. Anal. Appl. 302 (2005) 196–200.

54. D.S. Cîmpean, D. Popa, On the stability of the linear differential equation of higher order with constant coefficients, Appl. Math. Comput. 217 (2010) 4141–4146.

55. D.S. Cîmpean, D. Popa, Hyers-Ulam stability of Euler's equation, Appl. Math. Lett. 24 (2011) 1539–1543.

56. S. Czerwik, On the stability of the quadratic mapping in normed spaces, Abh. Math. Sem. Univ. Hamburg 62 (1992) 59–64.

57. G. Dolinar, Generalized stability of isometries, J. Math. Anal. Appl. 242 (2000) 39–56.

58. Iz. EL-Fassi, Generalized hyperstability of a Drygas functional equation on a restricted domain using Brzdęk's fixed point theorem, J. Fixed Point Theory Appl. (2017). doi:10.1007/s11784-017-0439-8

59. Iz. EL-Fassi, S. Kabbaj, On the hyperstability of a Cauchy-Jensen type functional equation in Banach spaces, Proyecciones J. Math. 34 (2015) 359–375.

60. Iz. EL-Fassi, S. Kabbaj, A. Charifi, Hyperstability of Cauchy-Jensen functional equations, Indagat. Math. 27 (2016) 855–867.

61. W. Fechner, Stability of a composite functional equation related to idempotent mappings, J. Approx. Theory 163 (2011) 328–335.

62. J.W. Fickett, Approximate isometries on bounded sets with an application to measure theory, Studia Math. 72 (1982) 37–46.

63. G.L. Forti, Hyers-Ulam stability of functional equations in several variables, Aequationes Math. 50 (1995) 143–190.

64. M. Gachpazan, O. Baghani, Hyers-Ulam stability of nonlinear integral equation, Fixed Point Theory Appl. 2010 (2010), Art. ID 927640, 6 pp.

65. M. Gachpazan, O. Baghani, Hyers-Ulam stability of Volterra integral equation, Int. J. Nonlinear Anal. Appl. 1 (2010) 19–25.

66. Z. Gajda, On stability of additive mappings, Int. J. Math. Math. Sci. 14 (1991) 431–434.

67. L. Găvruța, Matkowski contractions and Hyers-Ulam stability, Bul. Ştiinţ. Univ. Politeh. Timiş. Ser. Mat. Fiz. 53(67) (2008) 32–35.

68. P. Găvruța, A generalization of the Hyers-Ulam-Rassias stability of approximately additive mappings, J. Math. Anal. Appl. 184 (1994) 431–436.

69. P. Găvruța, L. Găvruța, A new method for the generalized Hyers-Ulam-Rassias stability, Int. J. Nonlinear Anal. Appl 1 (2010) 11–18.

70. R. Ger, Superstability is not natural, Rocznik Nauk.-Dydakt. Prace Mat. 159 (1993) 109–123.

71. J. Gevirtz, Stability of isometries on Banach spaces, Proc. Amer. Math. Soc. 89 (1983) 633–636.

72. P.M. Gruber, Stability of isometries, Trans. Amer. Math. Soc. 245 (1978) 263–277.

73. D.H. Hyers, On the stability of the linear functional equation, Proc. Nat. Acad. Sci. USA 27 (1941) 222–224.

74. D.H. Hyers, G. Isac, Th.M. Rassias, Stability of Functional Equations in Several Variables, Birkhäuser Boston, Inc., Boston, MA, 1998.

75. D.H. Hyers, Th.M. Rassias, Approximate homomorphisms, Aequationes Math. 44 (1992) 125–153.

76. D.H. Hyers, S.M. Ulam, On approximate isometries, Bull. Amer. Math. Soc. 51 (1945) 288–292.

77. D.H. Hyers, S.M. Ulam, Approximate isometries of the space of continuous functions, Ann. Math. 48 (1947) 285–289.

78. D.H. Hyers, S.M. Ulam, Approximately convex functions, Proc. Amer. Math. Soc. 3 (1952) 821–828.

79. K.W. Jun, Y.H. Lee, On the Hyers-Ulam-Rassias stability of a pexiderized quadratic inequality, Math. Inequal. Appl. 4 (2001) 93–118.

80. S.M. Jung, On the Hyers-Ulam stability of the functional equations that have the quadratic property, J. Math. Anal. Appl. 222 (1998) 126–137.

81. S.M. Jung, Hyers-Ulam-Rassias Stability of Functional Equations in Mathematical Analysis, Hadronic Press, Palm Harbor, 2001.

82. S.M. Jung, Hyers-Ulam stability of linear differential equation of first order, III, J. Math. Anal. Appl. 311 (2005) 139–146.

83. S.M. Jung, Hyers-Ulam stability of linear differential equations of first order, II, Appl. Math. Lett. 19 (2006) 854–858.

84. S.M. Jung, Hyers-Ulam stability of a system of first order linear differential equations with constant coefficients, J. Math. Anal. Appl. 320 (2006) 549–561.

85. S.M. Jung, A fixed point approach to the stability of a Volterra integral equation, Fixed Point Theory Appl. 2007, Art. ID 57064, 9 pp.

86. S.M. Jung, Hyers-Ulam-Rassias Stability of Functional Equations in Nonlinear Analysis, Springer, New York, 2011.

87. S.M. Jung, B. Kim, Stability of isometries on restricted domains, J. Korean Math. Soc. 37 (2000) 125–137.

88. S.M. Jung, D. Popa, Th.M. Rassias, On the stability of the linear functional equation in a single variable on complete metric groups, J. Glob. Optim. 59 (2014) 165–171.

89. S.M. Jung, M.Th. Rassias, A linear functional equation of third order associated to the Fibonacci numbers, Abstr. Appl. Anal. 2014 (2014), Art. ID 137468, 7 pp.

90. S.M. Jung, M.Th. Rassias, C. Mortici, On a functional equation of trigonometric type, Appl. Math. Comput. 252 (2015) 294–303.

91. S.M. Jung, H. Şevli, Power series method and approximate linear differential equations of second order, Adv. Difference Equ. 2013 (2013) 1–9.

92. Pl. Kannappan, The functional equation $f(xy) + f(xy^{-1}) = 2f(x)f(y)$ for groups, Proc. Amer. Math. Soc. 19 (1968) 69–74.

93. Z. Kominek, On a local stability of the Jensen functional equation, Demonstratio Math. 22 (1989) 499–507.

94. Y.H. Lee, On the stability of the monomial functional equation, Bull. Korean Math. Soc. 45 (2008) 397–403.

95. Y.H. Lee, K.W. Jun, A generalization of the Hyers-Ulam-Rassias stability of Jensen's equation, J. Math. Anal. Appl. 238 (1999) 305–315.

96. Y.H. Lee, K.W. Jun, A generalization of the Hyers-Ulam-Rassias stability of the Pexider equation, J. Math. Anal. Appl. 246 (2000) 627–638.

97. Y.H. Lee, S.M. Jung, M.Th. Rassias, On an n-dimensional mixed type additive and quadratic functional equation, Appl. Math. Comput. 228 (2014), 13–16.

98. Y. Li, Y. Shen, Hyers-Ulam stability of linear differential equations of second order, Appl. Math. Lett. 23 (2010) 306–309.

99. J. Lindenstrauss, A. Szankowski, Nonlinear perturbations of isometries, Astérisque 131 (1985) 357–371.

100. N. Lungu, D. Popa, Hyers-Ulam stability of a first order partial differential equation, J. Math. Anal. Appl. 385 (2012) 86–91.

101. N. Lungu, D. Popa, On the Hyers-Ulam stability of a first order partial differential equation, Carpathian J. Math. 28 (2012) 77–82.

102. Gy. Maksa, Zs. Páles, Hyperstability of a class of linear functional equations, Acta Math. Acad. Paedag. Nyiregyháziensis 17 (2001) 107–112.

103. T. Miura, G. Hirasawa, S.E. Takahasi, T. Hayata, A note on the stability of an integral equation, in: (Th.M. Rasssias, J. Brzdęk, eds.) Functional Equations in Mathematical Analysis, pp. 207–222, Springer Optimization and its Applications, vol. 52, Springer, New York-Dordrecht-Heidelberg-London, 2012.

104. T. Miura, S. Miyajima, S.E. Takahasi, A characterization of Hyers-Ulam stability of first order linear

differential operators, J. Math. Anal. Appl. 286 (2003) 136–146.

105. T. Miura, S. Miyajima, S.E. Takahasi, Hyers-Ulam stability of linear differential operator with constant coefficients, Math. Nachr. 258 (2003) 90–96.

106. J.R. Morales, E.M. Rojas, Hyers-Ulam and Hyers-Ulam-Rassias stability of nonlinear integral equations with delay, Int. J. Nonlinear Anal. Appl. 2 (2011) 1–6.

107. C. Mortici, M.Th. Rassias, S.M. Jung, On the stability of a functional equation associated with the Fibonacci numbers, Abstr. Appl. Anal. 2014 (2014), Art. ID 546046, 6 pp.

108. Z. Moszner, Sur les définitions différentes de la stabilité des équations fonctionnelles, Aequationes Math. 68 (2004) 260–274.

109. Z. Moszner, On the stability of functional equations, Aequationes Math. 77 (2009) 33–88.

110. Z. Moszner, On stability of some functional equations and topology of their target spaces, Ann. Univ. Paedagog. Crac. Stud. Math. 11 (2012) 69–94.

111. Z. Moszner, Stability has many names, Aequationes Math. 90 (2016) 983–999.

112. A. Najdecki, On stability of a functional equation connected with the Reynolds operator, J. Inequal. Appl. 2007, Art. ID 79816, 3 pp.

113. M. Obłoza, Hyers stability of the linear differential equation, Rocznik Nauk.-Dydakt. Prace Mat. 13 (1993) 259–270.

114. M. Omladič, P. Šemrl, On nonlinear perturbations of isometries, Math. Ann. 303 (1995) 617–628.

115. M. Piszczek, On selections of set-valued inclusions in a single variable with applications to several variables, Results Math. 64 (2013), 1–12.

116. M. Piszczek, Remark on hyperstability of the general linear equation, Aequationes Math. 88 (2014) 163–168.

117. M. Piszczek, Hyperstability of the general linear functional equation, Bull. Korean Math. Soc. 52 (2015) 1827–1838.

118. M. Piszczek, J. Szczawińska, Hyperstability of the Drygas functional equation, J. Funct. Spaces Appl. 2013 (2013), Art. ID 912718, 4 pp.

119. G. Pólya, G. Szegö, Problems and theorems in analysis, vol. 1, Part One, Ch. 3, Problem 99. Springer-verlag, Berlin-Heidelberg-New York, 1972.

120. D. Popa, I. Raşa, On the Hyers-Ulam stability of the linear differential equation, J. Math. Anal. Appl. 381 (2011) 530–537.

121. D. Popa, I. Raşa, Hyers-Ulam stability of the linear differential operator with nonconstant coefficients, Appl. Math. Comput. 219 (2012) 1562–1568.

122. A. Prástaro, Th.M. Rassias, Ulam stability in geometry of PDE's, Nonlinear Funct. Anal. Appl. 8 (2003) 259–278.

123. B. Przebieracz, Recent developments in the translation equation and its stability, In: (J. Brzdęk, K. Ciepliński, Th. M. Rassias, eds.) Developments in Functional Equations and Related Topics, Springer Optimization and Its Applications, to appear.

124. J.M. Rassias, Solution of a quadratic stability Hyers-Ulam type problem, Ricerche Mat. 50 (2001) 9–17.

125. Th.M. Rassias, On the stability of the linear mapping in Banach spaces, Proc. Amer. Math. Soc. 72 (1978) 297–300.

126. Th.M. Rassias, The stability of mappings and related topics, in: Report on the 27th Internat. Symp. on Functional Equations, Aequationes Math. 39 (1990) 292–293. Problem 16, 2°. (Same Report, p. 309.)

127. Th.M. Rassias, On the stability of the quadratic functional equation and its applications, Stud. Universitatis Babeş-Bolyai Math. 43 (1998) 89–124.

128. Th.M. Rassias, Properties of isometric mappings, J. Math. Anal. Appl. 235 (1999) 108–121.

129. Th.M Rassias, On the stability of functional equations and a problem of Ulam, Acta Appl. Math. 62 (2000) 23–130.

130. Th.M. Rassias, Isometries and approximate isometries, Int. J. Math. Math. Sci. 25 (2001) 73–91.

131. Th.M. Rassias, P. Šemrl, On the behaviour of mappings which do not satisfy Hyers-Ulam stability, Proc. Amer. Math. Soc. 114 (1992) 989–993.

132. J. Rätz, On approximately additive mappings, in General Inequalities 2, pp. 233–251, (Oberwolfach

1978) Birkhauser, Basel, Boston, 1980.

133. H. Rezaei, S.M. Jung, Th.M. Rassias, Laplace transform and Hyers-Ulam stability of the linear differential equations, J. Math. Anal. Appl. 403 (2013) 244–251.

134. I.A. Rus, Ulam stability of ordinary differential equations, Stud. Univ. Babeş-Bolyai Math. 54 (2009) 125–133.

135. I.A. Rus, Remarks on Ulam stability of the operatorial equations, Fixed Point Theory 10 (2009) 305–320.

136. I.A. Rus, Gronwall lemma approach to the Hyers-Ulam-Rassias stability of an integral equation, In: Nonlinear Analysis and Variational Problems, pp. 147–152, Springer Optimization and Its Applications, vol. 35, Springer, New York, 2010.

137. I.A. Rus, N. Lungu, Ulam stability of a nonlinear hyperbolic partial differential equation, Carpathian J. Math. 24 (2008) 403–408.

138. F. Skof, Proprieta' locali e approssimazione di operatori, Rend. Sem. Mat. Fis. Milano 53 (1983) 113–129.

139. F. Skof, Sulle δ-isometrie negli spazi normati, Rend. Mat. Appl. 10 (1990) 853–866.

140. F. Sládek, P. Zlatoš, A local stability principle for continuous group homomorphisms in nonstandard setting, Aequationes Math. 89 (2015) 991–1001.

141. R.L. Swain, Approximate isometries in bounded spaces, Proc. Amer. Math. Soc. 2 (1951) 727–729.

142. S.E. Takahasi, T. Miura, S. Miyajima, On the Hyers-Ulam stability of the Banach space-valued differential equation $y' = \lambda y$, Bull. Korean Math. Soc. 39 (2002) 309–315.

143. S.E. Takahasi, H. Takagi, T. Miura, S. Miyajima, The Hyers-Ulam stability constants of first order linear differential operators, J. Math. Anal. Appl. 296 (2004) 403–409.

144. S.M. Ulam, A Collection of Mathematical Problems, Interscience, New York, 1960.

145. G. Wang, M. Zhou, L. Sun, Hyers-Ulam stability of linear differential equations of first order, Appl. Math. Lett. 21 (2008) 1024–1028.

146. J.R. Wang, M. Fečkan, Practical Ulam-Hyers-Rassias stability for nonlinear equations, Math. Bohemica 142 (2017) 47–56.

147. J.R. Wang, M. Fečkan, Y. Zhou, Ulam's type stability of impulsive ordinary differential equations, J. Math. Anal. Appl. 395 (2012) 258–264.

148. J.R. Wang, Y. Zhou, M. Fečkan, Nonlinear impulsive problems for fractional differential equations and Ulam stability, Comp. Math. Appl. 64 (2012) 3389–3405.

149. D. Zhang, On hyperstability of generalised linear functional equations in several variables, Bull. Aust. Math. Soc. 92 (2015) 259–267.

CHAPTER 2

Ulam stability of operators in normed spaces

Contents

Abstract

In this chapter we suggest a new approach to the issue of Ulam stability. The majority of the existing results on Ulam stability are given for equations or operators acting on normed or metric spaces. We consider linear spaces endowed with gauges and investigate Ulam stability of linear operators acting on such spaces.

In this way we give a very general characterization for the Ulam stability of linear operators that is applied to the study of stability of some differential operators and some classical operators in approximation theory.

1. Introduction

It seems that the Ulam (often also called Hyers-Ulam) stability of linear operators was considered for the first time in [16, 17, 27], where a characterization has been obtained of such stability and a representation of the corresponding constants for linear operators. For the linear differential operator with constant coefficients in a Banach space, the authors of [27] proved that it is stable in this sense if and only if the characteristic equation has no pure imaginary solutions. Similar results are obtained in

[40] for weighted composition operators on $C(X)$, where X is a compact Hausdorff space. The stability of a linear composition operator of the second order was considered by J. Brzdęk and S.M. Jung in [7]. Next, Popa and Raşa obtained results on Ulam stability of some classical operators in approximation theory (Bernstein, Stancu, Szász-Mirakjan, Kantorovich, Beta, and others) and on their best constant [33, 34]. A new approach on Ulam stability of linear operators acting on linear spaces endowed with gauges has been considered by Brzdęk, Popa, and Raşa in [8], where, as applications, some results are given on the stability of the linear differential operators with constant coefficients with respect to different gauges.

2. Ulam stability with respect to gauges

Linear spaces endowed with gauges provide a general framework for investigating the Ulam stability of linear operators.

Let's start with the definition of a gauge; see also [16, 27].

Let $\mathbb{K}$ be one of the fields $\mathbb{R}$ or $\mathbb{C}$, and A a linear space over $\mathbb{K}$.

Definition 1. A function $\rho_A : A \to [0, +\infty]$ is called a semigauge on A if $\rho_A(0) = 0$ and $\rho_A(\lambda x) = |\lambda| \rho_A(x)$ for all $x \in A$, $\lambda \in \mathbb{K}$, $\lambda \neq 0$.

Definition 2. A semigauge for which $\rho_A(x) = 0$ if and only if $x = 0$ is called a gauge on A.

Many examples of semigauges and gauges will be displayed throughout the book. For the moment, let's mention a simple but universal example.

Example 2. The function

$$\sigma_A : A \to [0, +\infty], \qquad \sigma_A(x) = \begin{cases} 0, & x = 0, \\ +\infty, & x \neq 0, \end{cases}$$

is a gauge on A.

Now, let N be a linear subspace of A. As usual, the coset of an element $x \in A$ with respect to N is denoted by

$$\widetilde{x} := x + N = \{x + z \mid z \in N\}.$$

Then the quotient space is

$$A/N := \{\widetilde{x} \mid x \in A\}.$$

Proposition 1. *Let ρ_A be a semigauge on A. Then the function $\widetilde{\rho}_A : A/N \to [0, +\infty]$,*

$$\widetilde{\rho}_A(\widetilde{x}) := \inf_{z \in N} \rho_A(x - z) = \inf_{y \in \widetilde{x}} \rho_A(y), \qquad \widetilde{x} \in A/N,$$

is a semigauge on A/N.

Proof. Clearly

$$\widetilde{\rho}_A(\widetilde{0}) = \inf_{z \in N} \rho_A(0 - z) = \rho_A(0) = 0.$$

Now let $x \in A$, $\lambda \in \mathbb{K}$, $\lambda \neq 0$. Then

$$\widetilde{\rho}_A(\lambda \widetilde{x}) = \widetilde{\rho}_A(\widetilde{\lambda x}) = \inf_{u \in N} \rho_A(\lambda x - u) = \inf_{u \in N} \rho_A(\lambda x - \lambda u)$$

$$= \inf_{u \in N} |\lambda| \rho_A(x - u) = |\lambda| \widetilde{\rho}_A(\widetilde{x}).$$

This completes the proof. $\qquad\qquad\qquad\qquad\qquad\qquad\qquad\qquad\qquad\qquad\qquad\qquad$ $\square$

Example 3. Let $(A, \|\cdot\|)$ be a normed space and N be a dense linear subspace of A. Then $\rho_A := \|\cdot\|$ is a gauge, but $\widetilde{\rho}_A \equiv 0$ is only a semigauge on A/N.

Let A, B be linear spaces over $\mathbb{K}$, and ρ_A, ρ_B be semigauges on A and B, respectively. Let $L : A \to B$ be a linear operator and $N := \ker L$.

Definition 3. We say that L is Ulam stable with constant $K \geq 0$ if for each $x \in A$ such that $\rho_B(Lx) \leq 1$ there exists $z \in N$ with $\rho_A(x - z) \leq K$.

Consider the set

$$\mathfrak{K}(L) := \{K \geq 0 : L \text{ is Ulam stable with constant } K\},$$

and let $K(L) := \inf \mathfrak{K}(L)$.

Definition 4. The operator L is called Ulam stable if $\mathfrak{K}(L) \neq \emptyset$, i.e.,

$$K(L) < +\infty.$$

Remark 1. Clearly $\mathfrak{K}(L)$ is a subinterval of $[0, +\infty)$. The following example exhibits a nonzero operator L, for which $\mathfrak{K}(L) = [0, +\infty)$.

Example 4. Let $\mathbb{T}_n$ be the linear space of all polynomial functions of degree $\leq n$, defined on $\mathbb{K}$. Let $L : \mathbb{T}_1 \to \mathbb{T}_0$, $Lp = p'$, $p \in \mathbb{T}_1$. On $\mathbb{T}_1$ consider the semigauge $\rho_1(p) = |p(0)|$, $p \in \mathbb{T}_1$, and let ρ_0 be an arbitrary semigauge on $\mathbb{T}_0$.

Let $p \in \mathbb{T}_1$, $\rho_0(Lp) \leq 1$, and let $q \equiv p(0)$. Then $q \in \ker L$ and

$$\rho_1(p - q) = |(p - q)(0)| = 0,$$

which shows that $L \neq 0$ is Ulam stable with constant $K = 0$.

Now, let (A, ρ_A), (B, ρ_B), (C, ρ_C) be linear spaces endowed with semigauges. Let $V : A \to B$, $U : B \to C$ be linear operators, Ulam stable with constants $K \geq 0$ respectively, $J > 0$.

Theorem 37. *If $\ker U$ is contained in the range $R(V)$ of V, then $UV : A \to C$ is Ulam stable with the constant JK.*

Proof. Let $x \in A$ with $\rho_C(UVx) \leq 1$. Set $y := Vx$. Then $y \in B$ and $\rho_C(Uy) \leq 1$. Consequently, there exists $b \in \ker U$ such that

$$\rho_B(y - b) \leq J.$$

Since $b \in \ker U$, we have also $b \in R(V)$, i.e., there exists $a \in A$ with $b = Va$. Now $\rho_B(Vx - Va) \leq J$, i.e.,

$$\rho_B\left(V\frac{x - a}{J}\right) \leq 1.$$

This implies the existence of $u \in \ker V$ such that

$$\rho_A\left(\frac{x - a}{J} - u\right) \leq K.$$

Thus we have

$$\rho_A(x - (a + Ju)) \leq JK.$$

But $UV(a + Ju) = Ub = 0$, i.e.,

$$a + Ju \in \ker UV,$$

which means that UV is Ulam stable with constant JK.   □

Let again A, B be linear spaces with semigauges ρ_A, ρ_B, and $L : A \to B$ be a linear operator. Consider the set

$$\mathfrak{B}(L) := \{H \geq 0 : \rho_B(Lx) \leq H\rho_A(x), \ x \in A\},$$

and let $\rho(L) := \inf \mathfrak{B}(L)$.

Definition 5. We say that L is bounded if $\mathfrak{B}(L) \neq \varnothing$, i.e., $\rho(L) < +\infty$.

Remark 2. Clearly $\mathfrak{B}(L)$ is a subinterval of $[0, +\infty)$. The next example presents an operator $L \neq 0$ for which $\mathfrak{B}(L) = (0, +\infty)$.

Example 5. With notation from Example 2, let $L : (A, \sigma_A) \to (B, \sigma_B)$, $L \neq 0$. Then $\sigma_B(Lx) \leq H\sigma_A(x)$ for all $x \in A$ and all $H > 0$, i.e., $\mathfrak{B}(L) = (0, +\infty)$.

Proposition 2. *Let L be bounded and $x \in A$. If $\rho(L) > 0$ or $\rho_A(x) < +\infty$, then*

$$\rho_B(Lx) \leq \rho(L)\rho_A(x). \tag{2.1}$$

Proof. 1) Let $\rho_A(x) = 0$ and let $H \in \mathfrak{B}(L)$. Then $\rho_B(Lx) \leq H\rho_A(x)$ implies $\rho_B(Lx) = 0$, so (2.1) is satisfied.

2) Let $0 < \rho_A(x) < +\infty$. Then

$$\rho_B\left(\frac{Lx}{\rho_A(x)}\right) \leq H$$

for all $H \in \mathfrak{B}(L)$, which yields

$$\rho_B\left(\frac{Lx}{\rho_A(x)}\right) \leq \inf \mathfrak{B}(L) = \rho(L).$$

This entails (2.1).

3) If $\rho_A(x) = +\infty$, then by hypothesis $\rho(L) > 0$, and (2.1) is trivially satisfied. $\square$

Theorem 38. *Let $V : (A, \rho_A) \to (B, \rho_B)$ and $U : (B, \rho_B) \to (C, \rho_C)$ be bounded linear operators. Then $UV : (A, \rho_A) \to (C, \rho_C)$ is bounded, and*

$$\rho(UV) \leq \rho(U)\rho(V).$$

Proof. Let $\varepsilon > 0$ be given. There exist $0 < G < \rho(U) + \varepsilon$ and $0 < H < \rho(V) + \varepsilon$ such that

$$\rho_C(Uy) \leq G\rho_B(y), \qquad y \in B,$$

$$\rho_B(Vz) \leq H\rho_A(z), \qquad z \in A.$$

Let $x \in A$ be given. Then $Vx \in B$ and

$$\rho_C(UVx) \leq G\rho_B(Vx) \leq GH\rho_A(x) \leq (\rho(U) + \varepsilon)(\rho(V) + \varepsilon)\rho_A(x).$$

It follows that UV is bounded and

$$\rho(UV) \leq (\rho(U) + \varepsilon)(\rho(V) + \varepsilon)$$

for all $\varepsilon > 0$. Letting $\varepsilon \to 0$ we get

$$\rho(UV) \le \rho(U)\rho(V),$$

and this concludes the proof. $\square$

Again, let A, B be linear spaces with semigauges ρ_A, ρ_B, and let $L : A \to B$ be a linear operator. Let $N = \ker L$, and $R(L) \subset B$ be the range of L. Consider the linear operator $\widetilde{L} : A/N \to R(L)$ given by

$$\widetilde{L}\widetilde{x} := Lx, \quad \widetilde{x} \in A/N, \ x \in A.$$

Then $\widetilde{L}$ is bijective. Let $\widetilde{L}^{-1} : R(L) \to A/N$ be the inverse of $\widetilde{L}$.

Remembering that $\widetilde{\rho}_A : A/N \to [0, +\infty]$ is a semigauge on A/N, let us investigate the relationship between the Ulam stability of L and the boundedness of $\widetilde{L}^{-1} : (R(L), \rho_B) \to (A/N, \widetilde{\rho}_A)$.

Remember that

$$\rho(\widetilde{L}^{-1}) := \inf \left\{ H \ge 0 : \widetilde{\rho}_A(\widetilde{L}^{-1}y) \le H\rho_B(y), \ y \in R(L) \right\}.$$

Theorem 39. *If $\widetilde{L}^{-1}$ is bounded and $\delta > 0$, then L is Ulam stable with the constant $\rho(\widetilde{L}^{-1}) + \delta$.*

Proof. Fix $\delta > 0$. Let $x \in A$ with $\rho_B(Lx) \le 1$, and set $y = Lx$. Then $\rho_B(y) \le 1$ and

$$\widetilde{L}\widetilde{x} = Lx = y,$$

i.e., $\widetilde{L}^{-1}y = \widetilde{x}$. Let $H \in \mathfrak{B}(\widetilde{L}^{-1})$ such that $H < \rho(\widetilde{L}^{-1}) + \delta$. Consequently we have

$$\widetilde{\rho}_A(\widetilde{x}) = \widetilde{\rho}_A(\widetilde{L}^{-1}y) \le H\rho_B(y) \le H < \rho(\widetilde{L}^{-1}) + \delta.$$

Therefore

$$\inf_{u \in N} \rho_A(x - u) = \widetilde{\rho}_A(\widetilde{x}) < \rho(\widetilde{L}^{-1}) + \delta,$$

which entails the existence of $z \in N$ with

$$\rho_A(x - z) < \rho(\widetilde{L}^{-1}) + \delta.$$

This shows that L is Ulam stable with constant $\rho(\widetilde{L}^{-1}) + \delta$. $\square$

Theorem 40. *If L is Ulam stable with constant $K > 0$, then*

$$\widetilde{\rho}_A(\widetilde{L}^{-1}y) \le K\rho_B(y), \qquad y \in R(L).$$

Proof. Clearly the above inequality is satisfied if $\rho_B(y) = +\infty$. It remains to consider the case when $\rho_B(y) < +\infty$.

Take $x \in A$ with $y = Lx$. Let $\varepsilon > 0$. Then

$$\rho_B\left(L\frac{x}{\rho_B(y) + \varepsilon}\right) = \rho_B\left(\frac{y}{\rho_B(y) + \varepsilon}\right) < 1,$$

and therefore there exists $z \in N$ with

$$\rho_A\left(\frac{x}{\rho_B(y) + \varepsilon} - z\right) \le K.$$

This entails

$$\rho_A(x - (\rho_B(y) + \varepsilon)z) \le (\rho_B(y) + \varepsilon)K.$$

Since $(\rho_B(y) + \varepsilon)z \in N$, we conclude that

$$\widetilde{\rho}_A(\widetilde{x}) \le (\rho_B(y) + \varepsilon)K$$

for all $\varepsilon > 0$, i.e.,

$$\widetilde{\rho}_A(\widetilde{x}) \le K\rho_B(y).$$

On the other hand, $\widetilde{L}\widetilde{x} = Lx = y$, so that $\widetilde{x} = \widetilde{L}^{-1}y$. Consequently,

$$\widetilde{\rho}_A(\widetilde{L}^{-1}y) \le K\rho_B(y),$$

and this concludes the proof. □

Theorem 41. *Let $L : (A, \rho_A) \to (B, \rho_B)$ be a linear operator. Then:*
(i) $K(L) = \rho(\widetilde{L}^{-1})$.
(ii) L is Ulam stable if and only if $\widetilde{L}^{-1}$ is bounded.

Proof. According to Theorem 40,

$$\Re(L) \setminus \{0\} \subset \mathfrak{B}(\widetilde{L}^{-1}).$$

Since $\Re(L)$ is a subinterval of $[0, +\infty)$, we deduce that

$$\inf \mathfrak{B}(\widetilde{L}^{-1}) \le \inf(\Re(L) \setminus \{0\}) = \inf \Re(L).$$

Therefore,

$$\rho(\widetilde{L}^{-1}) \le K(L) \le +\infty. \tag{2.2}$$

If $\rho(\widetilde{L}^{-1}) = +\infty$, then (i) is a consequence of (2.2). Suppose that $\rho(\widetilde{L}^{-1}) < +\infty$. Then, by Theorem 39, $K(L) \le \rho(\widetilde{L}^{-1}) + \delta$ for each $\delta > 0$, whence $K(L) \le \rho(\widetilde{L}^{-1})$. Combined with (2.2), this proves (i). Obviously, (ii) is a consequence of (i), and the proof is finished. □

Example 6. Let $(A, \|\cdot\|)$ and $(B, \|\cdot\|)$ be normed spaces. Let $L : A \to B$ be a linear operator, $L \neq 0$, such that $\ker L$ is dense in A. Then $\mathfrak{B}(\widetilde{L}^{-1}) = [0, +\infty)$ and $\mathfrak{K}(L) = (0, +\infty)$.

3. Closed operators

Let X and Y be normed spaces and $L : D(L) \to Y$ be a linear operator, where the domain $D(L)$ of L is a linear subspace of X. Suppose that L is a closed operator, i.e., its graph

$$G(L) := \{(x, y) \mid x \in D(L), y = Lx\}$$

is closed in the normed space $X \times Y$ with the norm $\|(x, y)\| = \|x\| + \|y\|$. (For definitions, see, e.g., [21]).

Then $N(L)$, the kernel of L, is a closed subspace of X, and $X/N(L)$ is a normed space with the norm given by

$$\|\bar{x}\| := \inf \{\|x - u\| \, | \, u \in N(L)\}, \qquad \widetilde{x} \in X/N(L).$$

In this context, according to Definition 3, L is Ulam stable with constant $K \geq 0$ if and only if one of the following equivalent conditions is satisfied:

(i)   For each $x \in D(L)$ such that $\|Lx\| \leq 1$ there exists $z \in N(L)$ with

$$\|x - z\| \leq K.$$

(ii)   For each $x \in D(L)$ and $\varepsilon > 0$ such that $\|Lx\| \leq \varepsilon$ there exists $u \in N(L)$ with

$$\|x - u\| \leq K\varepsilon.$$

(iii)   For each $x \in D(L)$ there exists $u \in N(A)$ such that

$$\|x - u\| \leq K \|Lx\|.$$

See also [17, Remark 2.1]. As in Section 1, let

$$K(L) := \inf \{K \geq 0 : L \text{ is Ulam stable with constant } K\}.$$

Consider also the bijective operator $\widetilde{L}^{-1} : R(L) \to X/N(L)$. By using Theorem 41 we can state the following:

Theorem 42. *L is Ulam stable if and only if $\widetilde{L}^{-1}$ is bounded. Moreover,*

$$K(L) = \left\| \widetilde{L}^{-1} \right\|.$$

The Ulam stability of L can be related also to the closedness of its range $R(L)$. More precisely, we have the next theorem.

Theorem 43. *Let X, Y be Banach spaces and $L : D(L) \subset X \to Y$ be a closed operator. The following statements are equivalent:*

(a) L is Ulam stable;
(b) $\widetilde{L}^{-1}$ is bounded;
(c) R(L) is closed.

If one of them is true, then

$$K(L) = \left\| \widetilde{L}^{-1} \right\|.$$

Proof. $(a) \Leftrightarrow (b)$: see Theorem 3.1. $(b) \Leftrightarrow (c)$: see [21, Theorem 4.5.2, p. 231]. □

Remark 3. Let $L : X \to Y$ be a bounded linear operator and X and Y be Banach spaces. Then L is closed and Theorem 42 can be applied to it. The corresponding result can be found in [40, Theorem 2]. The Ulam stability of closed operators acting between Hilbert spaces is investigated in [17, 18, 19, 30].

Again, let X and Y be Banach spaces and $L : D(L) \subset X \to Y$ a closed operator. Suppose that there exists a closed linear subspace M of X such that the restriction $L_0 : D(L) \cap M \to R(L)$ of L (i.e., $L_0 x = Lx$ for all $x \in D(L) \cap M$) is invertible. Since L is closed, it is easy to see that L_0 is closed, and therefore $L_0^{-1} : R(L) \to D(L) \cap M$ is closed. Under these assumptions we have the following:

Theorem 44. *(a) L_0^{-1} is bounded if and only if $R(L)$ is closed.*
 (b) If $R(L)$ is closed, then L is Ulam stable with constant $\left\| L_0^{-1} \right\|$.

Proof. (a) If L_0^{-1} is bounded, then $R(L)$ is closed; apply, e.g., [23, Lemmna 4.13-5]. If $R(L)$ is closed, the Closed Graph Theorem (see, e.g., [23, 4.13-2]) shows that L_0^{-1} is bounded.
 (b) Let $x \in D(L)$ and $y := Lx$, $z := L_0^{-1} y$. Then $z \in D(L) \cap M$ and

$$Lz = L(L_0^{-1} y) = y = Lx.$$

Let $u := x - z$. We have $u \in N(L)$ and

$$\|x - u\| = \|z\| = \left\| L_0^{-1} y \right\| \le \left\| L_0^{-1} \right\| \|y\|,$$

so that

$$\|x - u\| \le \left\| L_0^{-1} \right\| \|Lx\|.$$

This shows that L is Ulam stable with constant $\left\| L_0^{-1} \right\|$. □

Remark 4. If X and Y are Hilbert spaces, results related to Theorem 44 can be found in [17].

Example 7. Here we present an example in which Theorem 44 can be applied.

Let X be a Banach space and $(T(t))_{t \geq 0}$ be a C_0-semigroup of operators on X with infinitesimal generator $A : D(A) \subset X \to X$. Then A is a closed linear operator and $D(A)$ is dense in X. Suppose that for each $x \in X$ there exists the limit

$$Tx := \lim_{t \to \infty} T(t)x.$$

Examples of such semigroups can be found in [3] and the references therein.

It is known (see [13, Chapter 1, 9.14, 9.15] and [41, Chapter 12]) that T is a bounded linear projection and the operator $A_0 : D(A) \cap N(T) \to R(A)$, $A_0 x := Ax$, is invertible; its inverse is $-V$, where V is the potential operator

$$Vx = \int_0^\infty T(t)x \, dt, \qquad x \in R(A).$$

So, Theorem 44 can be applied whenever $R(A)$ is closed. Generators A with this property are studied, e.g., in [22, 25] and the references given there. The simplest example is concerned with $X = C[0, 1]$, the space of all continuous real-valued functions defined on $[0, 1]$, endowed with the supremum norm. The infinitesimal generator is $A : D(A) \to C[0, 1]$, where

$$D(A) := \left\{ f \in C[0, 1] \cap C^2(0, 1) : \lim_{x \to 0+} x(1 - x)f''(x) = \lim_{x \to 1-} x(1 - x)f''(x) = 0 \right\}$$

and, for $f \in D(A)$, $x \in [0, 1]$,

$$Af(x) := \begin{cases} 0, & x \in \{0, 1\}, \\ \frac{x(1-x)}{2} f''(x), & 0 < x < 1. \end{cases}$$

According to [2, Theorem 6.3.5], this is the infinitesimal generator of the semigroup described in [2, (6.3.64)]. The projection $T : C[0, 1] \to C[0, 1]$ is given by (see [2, (6.2.20), (6.3.19)])

$$Tf(x) = (1 - x)f(0) + xf(1), \qquad f \in C[0, 1], \ x \in [0, 1].$$

Obviously,

$$N(T) = \{g \in C[0, 1] : g(0) = g(1) = 0\}.$$

For $f \in C[0, 1]$ and $0 \leq x \leq 1$, let

$$Wf(x) := 2\left((1 - x) \int_0^x \frac{f(t)}{1 - t} dt + x \int_x^1 \frac{f(t)}{t} dt \right).$$

It is a matter of calculus to prove that

$$\lim_{x \to 0^+} Wf(x) = \lim_{x \to 1^-} Wf(x) = 0, \qquad f \in C[0, 1].$$

Thus W can be considered as a positive linear operator $W : C[0, 1] \to C[0, 1]$. If we denote by $\mathbb{1}$ the function with constant value 1, then the norm of W acting on the Banach space $(C[0, 1], \|\cdot\|_\infty)$ is $\|W\| = \|W\mathbb{1}\|_\infty$. Since

$$W\mathbb{1}(x) = -2((1 - x)\log(1 - x) + x\log x),$$

we get $\|W\| = \log 4$.

It is easy to verify that if $g \in C[0, 1]$, $g(0) = g(1) = 0$, then $Wg \in D(A) \cap N(T)$ and $AWg = -g$. It follows immediately that

$$R(A) = \{g \in C[0, 1] : g(0) = g(1) = 0\} = N(T),$$

and if V is the restriction of W to $R(A)$,

$$V : R(A) \to D(A) \cap N(T), \quad Vg = Wg, \quad g \in R(A),$$

then $-V$ is the inverse of the operator

$$A_0 : D(A) \cap N(T) \to R(A), \quad A_0 f = Af, \quad f \in D(A) \cap N(T).$$

In order to find the norm of V, let us remark first that if $g \in R(A)$, $\|g\|_\infty \leq 1$, then

$$\|Vg\|_\infty = \|Wg\|_\infty \leq \|W\| \cdot \|g\|_\infty \leq \log 4.$$

Thus $\|V\| \leq \log 4$. On the other hand, choose $g_n \in R(A)$, $0 \leq g_n \leq 1$, such that

$$\lim_{n \to \infty} g_n(x) = 1, \qquad 0 < x < 1.$$

Then $\|g_n\|_\infty \leq 1$ and

$$\|Vg_n\|_\infty = \|Wg_n\|_\infty \leq \log 4.$$

Moreover,

$$\|Vg_n\|_\infty \geq Vg_n(\tfrac{1}{2}) = \int_0^{\frac{1}{2}} \frac{g_n(t)}{1 - t} dt + \int_{\frac{1}{2}}^1 \frac{g_n(t)}{t} dt$$

$$\to \int_0^{\frac{1}{2}} \frac{1}{1 - t} dt + \int_{\frac{1}{2}}^1 \frac{1}{t} dt = \log 4.$$

We conclude that

$$\|Vg_n\|_\infty \to \log 4,$$

which shows that $\|V\| \geq \log 4$.

To resume, we have $\|V\| = \log 4$, and now Theorem 44 tells us that A is Ulam stable with constant $\log 4$.

Theorem 45. *Let X be a Banach space, $A : D(A) \subset X \to X$ be a dissipative linear operator, and I be the identity operator on X. Then $\lambda I - A$ is Ulam stable with constant $\frac{1}{\lambda}$, for each $\lambda > 0$.*

Proof. According to [31, Theorem 1.4.2], A is dissipative if and only if

$$\|(\lambda I - A)x\| \geq \lambda \|x\|, \text{ for all } x \in D(A) \text{ and } \lambda > 0.$$

So, if A is dissipative, $x \in D(A)$ and $\lambda > 0$, then

$$\|x - 0\| \leq \frac{1}{\lambda} \|(\lambda I - A)x\|,$$

which shows that $\lambda I - A$ is Ulam stable with constant $\frac{1}{\lambda}$. $\square$

In order to state the next result, let X and Y be normed spaces, and $A : D(A) \subset X \to Y$ be a linear operator. Let us endow $D(A)$ with the graph norm, defined as

$$\|x\|_G := \|x\| + \|Ax\|, \qquad x \in D(A).$$

Consider the operator

$$A_G := (D(A), \|\cdot\|_G) \subset X \to Y, \qquad A_G x = Ax, \qquad x \in D(A).$$

Then

$$\|A_G x\| = \|Ax\| \leq \|x\| + \|Ax\| = \|x\|_G,$$

which shows that $\|A_G\| \leq 1$.

Theorem 46. *(a) A is Ulam stable with constant K if and only if A_G is Ulam stable with constant $K + 1$.*
(b) $\widetilde{A}^{-1}$ is bounded if and only if $\widetilde{A_G}^{-1}$ is bounded. Moreover,

$$\left\|\widetilde{A_G}^{-1}\right\| = \left\|\widetilde{A}^{-1}\right\| + 1.$$

Proof. Let $u \in N(A) = N(A_G)$. Then, for $x \in D(A)$,

$$\|x - u\| \leq K\|Ax\| \Leftrightarrow \|x - u\| + \|A(x - u)\| \leq (K + 1)\|Ax\|$$

$$\Leftrightarrow \|x - u\|_G \leq (K + 1)\|A_G x\|.$$

This proves the statement (a). Combining (a) with Theorem 41 we get (b). $\square$

Remark 5. When X and Y are Hilbert spaces, results related to Theorem 46 can be found in [17].

Theorem 47. *Let X, Y be normed spaces, $A : X \to Y$ be a bounded linear operator, and Z be a dense linear subspace of X. If the restriction $A : Z \to Y$ is Ulam stable with constant $K > 0$, then $A : X \to Y$ is Ulam stable with each constant $L > K$.*

Proof. Let $x \in X$, $\|Ax\| \le 1$. Let

$$\varepsilon = \frac{L - K}{2}, \qquad c = \frac{L + K}{2K}.$$

Then $\varepsilon > 0$, $c > 1$, $cK + \varepsilon = L$.

Let $z_n \in Z$, $z_n \to x$. Since A is bounded, we have $\|Az_n\| \to \|Ax\| < c$. Therefore, there exists $n \in \mathbb{N}$ (positive integers) such that $\|z_n - x\| \le \varepsilon$ and $\|Az_n\| \le c$. By hypothesis, $A : Z \to Y$ is Ulam stable with constant K, and consequently there exists $z_0 \in Z$, $Az_0 = 0$, $\|z_n - z_0\| \le Kc$. This yields

$$\|x - z_0\| \le \|x - z_n\| + \|z_n - z_0\| \le \varepsilon + Kc = L,$$

and so $A : X \to Y$ is Ulam stable with constant L. □

In [16] the authors investigated the circumstances when a linear operator L is Ulam stable with constant $K(L)$. Here is their result.

Theorem 48 ([16]). *Let A, B be linear spaces with semigauges ρ_A, ρ_B, and let $L : A \to B$ be an Ulam stable linear operator. The following two statements are equivalent:*

(i) *L is Ulam stable with constant $K(L)$.*
(ii) *For each $x \in A$ with $\widetilde{\rho}_A(\widetilde{x}) = K(L)$ and $\rho_B(Lx) \le 1$, there exists $u \in N(L)$ such that $\rho_A(x - u) = \widetilde{\rho}_A(\widetilde{x})$.*

Proof. Suppose that (i) holds. Let $x \in A$ with $\widetilde{\rho}_A(\widetilde{x}) = K(L)$ and $\rho_B(Lx) \le 1$. Then there exists $u \in N(L)$ such that $\rho_A(x - u) \le K(L)$. Therefore,

$$\widetilde{\rho}_A(\widetilde{x}) \le \rho_A(x - u) \le K(L) = \widetilde{\rho}_A(\widetilde{x}),$$

which implies $\rho_A(x - u) = \widetilde{\rho}_A(\widetilde{x})$, and so (ii) is verified.

Conversely, suppose that (ii) holds. Let $x \in A$ with $\rho_B(Lx) \le 1$. Let $K \in \mathfrak{R}(L)$, i.e., L is Ulam stable with constant K. Then there exists $u \in N(L)$ such that $\rho_A(x - u) \le K$. It follows that

$$\widetilde{\rho}_A(\widetilde{x}) \le K$$

for all $K \in \mathfrak{R}(L)$, and so

$$K(L) = \inf \mathfrak{R}(L) \geq \widetilde{\rho}_A(\widetilde{x}).$$

If $\widetilde{\rho}_A(\widetilde{x}) < K(L)$, there exists $u \in N(L)$ with $\rho_A(x - u) < K(L)$.
If $\widetilde{\rho}_A(\widetilde{x}) = K(L)$, (ii) shows that there exists $u \in N(L)$ with

$$\rho_A(x - u) = \widetilde{\rho}_A(\widetilde{x}) = K(L).$$

So, in both cases there exists $u \in N(L)$ such that $\rho_A(x - u) \leq K(L)$, which means that L is Ulam stable with constant $K(L)$. Thus (i) is verified, and this ends the proof.

$\square$

Corollary 1 ([16]). *Let $(A, \|\cdot\|)$ be a normed space and $L : (A, \|\cdot\|) \to (B, \rho_B)$ an Ulam stable linear operator. If $N(L)$ is a proximinal subspace of A, then L is Ulam stable with constant $K(L)$.*

Proof. By the definition of proximinality, for each $x \in A$ there exists $u \in N(L)$ such that

$$\|x - u\| = \inf \{\|x - z\| : z \in N(L)\}.$$

This validates the statement (ii) of Theorem 48, and so (i) is also valid. $\square$

Remark 6. In Example 6 we have an Ulam stable operator L, which is not stable with constant $K(L)$. Other examples of this kind can be found in [16, 34].

Theorem 49. *Let X be a normed space, $T : X \to X$ a compact linear operator, I the identity operator on X, and $\lambda \neq 0$. Then $A := T - \lambda I$ is Ulam stable.*

Proof. As usual, let $\widetilde{A} : X/N(A) \to R(A)$, $\widetilde{A}\widetilde{x} = Ax$, $x \in X$. We shall prove that $\widetilde{A}^{-1} : R(A) \to X/N(A)$ is bounded, where

$$\|\widetilde{x}\| := \inf \{\|x - u\| : u \in N(A)\}, \qquad \widetilde{x} \in X/N(A).$$

According to [23, Lemma 8.5-2], there exists $c > 0$ such that for each $y \in R(A)$ there is an $x \in X$ with $Ax = y$ and

$$\|x\| \leq c \|y\|.$$

Consequently,

$$\left\|\widetilde{A}^{-1}y\right\| = \|\widetilde{x}\| \leq \|x\| \leq c \|y\|,$$

i.e., $\left\|\widetilde{A}^{-1}y\right\| \leq c \|y\|$ for all $y \in R(A)$.

This shows that $\widetilde{A}^{-1}$ is bounded, and it suffices to apply Theorem 42 in order to conclude that A is Ulam stable.   □

4. Some differential operators on bounded intervals

This section is based on some results from [20, Chapter 11]. Let $a, b \in \mathbb{R}, a < b, n \geq 0$ and $w_k \in C^{n+1-k}[a, b], w_k(t) > 0$, for all $k = 0, 1, \ldots, n, t \in [a, b]$. Define, for $t \in [a, b]$,

$$u_0(t) := w_0(t),$$
$$u_1(t) := w_0(t) \int_a^t w_1(s_1)ds_1,$$
$$u_2(t) := w_0(t) \int_a^t w_1(s_1) \int_a^{s_1} w_2(s_2)ds_2ds_1,$$
$$\ldots$$
$$u_n(t) := w_0(t) \int_a^t w_1(s_1) \int_a^{s_1} w_2(s_2) \ldots \int_a^{s_{n-1}} w_n(s_n)ds_n \ldots ds_1.$$

Consider also the differential operators

$$D_j f := \frac{d}{dt}\frac{f}{w_j}, \qquad j = 0, 1, \ldots, n, f \in C^1[a, b],$$

and define

$$A := D_n \ldots D_1 D_0 : C^{n+1}[a, b] \to C[a, b].$$

Then the differential operator A satisfies

$$Au_j = 0, \qquad j = 0, 1, \ldots, n,$$

i.e.,

$$u_j \in N(A), \qquad j = 0, 1, \ldots, n.$$

In what follows we need the function

$$\varphi_n(t; x) := \begin{cases} \phi(t, x), & a \leq x \leq t \leq b; \\ 0, & a \leq t < x \leq b, \end{cases}$$

where

$$\phi(t, x) = w_0(t) \int_x^t w_1(s_1) \int_x^{s_1} w_2(s_2) \ldots \int_x^{s_{n-1}} w_n(s_n)ds_n \ldots ds_1.$$

Let

$$K := \int_a^b \varphi_n(b; x)dx.$$

To each function $f \in C^{n+1}[a, b]$ let's associate the coefficients

$$a_0(f) := \frac{f(a)}{w_0(a)}, \qquad a_j(f) := \frac{1}{w_j(a)}(D_{j-1} \ldots D_0 f)(a), \qquad j = 1, \ldots, n.$$

Then

$$p_f := \sum_{j=0}^{n} a_j(f)u_j \in N(A).$$

In Chapter 11 of their book [20], S. Karlin and W.J. Studden prove that

$$f(t) - p_f(t) = \int_a^t \varphi_n(t; x)(Af)(x)dx, \qquad t \in [a, b], f \in C^{n+1}[a, b]. \qquad (2.3)$$

If we denote by $\|\cdot\|_\infty$ the uniform norm on $C[a, b]$, then

$$\begin{aligned}
\|f - p_f\|_\infty &= \sup_{t \in [a,b]} |f(t) - p_f(t)| \\
&= \sup_{t \in [a,b]} \left| \int_a^t \varphi_n(t; x)(Af)(x)dx \right| \\
&\leq \sup_{t \in [a,b]} \left(\int_a^t \varphi_n(t; x)dx \right) \|Af\|_\infty \\
&= \left(\int_a^b \varphi_n(b; x)dx \right) \|Af\|_\infty = K \|Af\|_\infty.
\end{aligned}$$

So, we have the following:

Theorem 50. *With the above notation, the differential operator $A : C^{n+1}[a, b] \to C[a, b]$ is Ulam stable with constant*

$$K = \int_a^b \varphi_n(b; x)dx.$$

Example 8. Let $w_k(t) = 1$ for $t \in [a, b]$ and $k = 0, 1, \dots, n$. Then $D_j f = \frac{d}{dt}f$ for $j = 0, 1, \dots, n$, and

$$A : C^{n+1}[a, b] \to C[a, b], \qquad Af = f^{(n+1)}.$$

Moreover, $u_j(t) = \frac{(t-a)^j}{j!}$ for $j = 0, 1, \dots, n$,

$$\varphi_n(t; x) = \begin{cases} \frac{(t-x)^n}{n!}, & a \leq x \leq t \leq b, \\ 0, & a \leq t < x \leq b, \end{cases}$$

and $a_j(f) = f^{(j)}(a)$ for $j = 0, 1, \dots, n$. Consequently, p_f is Taylor's polynomial

$$p_f(t) = \sum_{j=0}^{n} \frac{f^{(j)}(a)}{j!}(t - a)^j,$$

and (2.3) becomes Taylor's formula:

$$f(t) - \sum_{j=0}^{n} \frac{f^{(j)}(a)}{j!}(t-a)^j = \int_{a}^{t} \frac{(t-x)^n}{n!} f^{(n+1)}(x)dx,$$

for all $f \in C^{n+1}[a,b]$ and $t \in [a,b]$.

In this case,

$$K = \int_{a}^{b} \varphi_n(b;x)dx = \int_{a}^{b} \frac{(b-x)^n}{n!}dx = \frac{(b-a)^{n+1}}{(n+1)!}.$$

We conclude that the operator $A = \left(\frac{d}{dt}\right)^{n+1}$ is Ulam stable on $C^{n+1}[a,b]$ with constant

$$\frac{(b-a)^{n+1}}{(n+1)!}.$$

Let us remark that the kernel $N(A)$ of this operator is Π_n, the space of all polynomial functions of degree $\leq n$ defined on $[a,b]$. It is well known that this subspace of $C^{n+1}[a,b]$ is proximinal, so that A is Ulam stable with constant $K(A)$; see Corollary 1. We are in a position to determine $K(A)$.

Proposition 3. *For the operator* $A = \left(\frac{d}{dt}\right)^{n+1} : C^{n+1}[a,b] \to C[a,b]$ *we have*

$$K(A) = \frac{(b-a)^{n+1}}{(n+1)!2^{2n+1}} \tag{2.4}$$

Proof. Consider the Chebyshev nodes

$$a_k := \frac{a+b}{2} + \frac{b-a}{2} \cos \frac{(2k+1)\pi}{2(n+1)}, \qquad k = 0, 1, \dots, n.$$

For $f \in C^{n+1}[a,b]$ let $L_n f \in \Pi_n$ be the Lagrange interpolation polynomial associated with f and the nodes $a_0, a_1, \dots, a_n$. It is well known that

$$\|f - L_n f\|_\infty \leq \frac{(b-a)^{n+1}}{(n+1)!2^{2n+1}} \left\|f^{(n+1)}\right\|_\infty.$$

Since $L_n f \in N(A)$, we see that A is Ulam stable with constant

$$\frac{(b-a)^{n+1}}{(n+1)!2^{2n+1}},$$

i.e.,

$$K(A) \le \frac{(b-a)^{n+1}}{(n+1)!2^{2n+1}}.$$ (2.5)

On the other hand, let K be an arbitrary Ulam constant for A. Then for each $f \in C^{n+1}[a,b]$ with $\|Af\|_\infty \le 1$ there exists $p \in \Pi_n$ such that $\|f - p\|_\infty \le K$. Take

$$f = \frac{e_{n+1}}{(n+1)!},$$

where $e_j(t) = t^j$ for $j = 0, 1, \ldots$. Then $Af = e_0$, so that $\|Af\|_\infty = 1$, and there exists $p \in \Pi_n$ with

$$\left\| \frac{e_{n+1}}{(n+1)!} - p \right\|_\infty \le K.$$

This yields

$$\|e_{n+1} - (n+1)!p\|_\infty \le (n+1)!K.$$

But it is known (see, e.g., [23, Section 6.4]) that

$$\frac{(b-a)^{n+1}}{2^{2n+1}} \le \|e_{n+1} - q\|_\infty, \qquad q \in \Pi_n.$$

We conclude that

$$\frac{(b-a)^{n+1}}{2^{2n+1}} \le (n+1)!K,$$

and therefore

$$\frac{(b-a)^{n+1}}{(n+1)!2^{2n+1}} \le K(A).$$ (2.6)

Now (2.4) is a consequence of (2.5) and (2.6). $\qquad\square$

5. Stability of the linear differential operator with respect to different norms

In this section we present some results obtained in [8].
Let $a, b \in \mathbb{R} \cup \{+\infty, -\infty\}$, $a < b$, and $I := (a, b)$. Let X be a Banach space over $\mathbb{K}$. Given $f \in C(I, X)$, $r \in (0, \infty)$, we write

$$\|f\|_r := \left(\int_I \|f(t)\|^r \, dt \right)^{1/r}, \qquad \|f\|_\infty := \sup \{\|f(t)\| : t \in I\}.$$

Then $\|\cdot\|_r$ and $\|\cdot\|_\infty$ are gauges on $C(I, X)$.
Let $\lambda \in \mathbb{K}$ and $D_\lambda : C^1(I, X) \to C(I, X)$ be defined by

$$D_\lambda f := f' + \lambda f, \qquad f \in C^1(I, X).$$

Then $N_\lambda := \ker D_\lambda$ is the subspace of $C^1(I, X)$ consisting of the functions $e^{-\lambda t}c$ $(t \in I)$ with $c \in X$. Moreover, the range of D_λ is $C(I, X)$ and

$$\widetilde{D}_\lambda^{-1}(y) = \left\{ f_k \in C^1(I, X) : k \in X \right\}, \qquad y \in C(I, X),$$

where

$$f_k(t) = e^{-\lambda t} \left(\int_{t_0}^t e^{\lambda s} y(s) ds + k \right), \qquad t \in I,$$

and t_0 is an arbitrarily fixed point in I. Clearly, as we have observed before, $\widetilde{D}_\lambda^{-1} :$ $C(I, X) \to C^1(I, X)/N_\lambda$ is bijective.

Let $p \in [1, +\infty]$, $r \in (0, \infty]$ and assume that the spaces $C^1(I, X)$ and $C(I, X)$ are endowed with the gauges $\|\cdot\|_r$ and $\|\cdot\|_p$, respectively. The associated semigauge on $C^1(I, X)/N_\lambda$ will be

$$\widetilde{\rho}_r(\widetilde{f}) := \inf_{g \in N_\lambda} \|f - g\|_r, \qquad f \in C^1(I, X). \tag{2.7}$$

Remark 7. According to Definition 5, the boundedness of $\widetilde{D}_\lambda^{-1}$ means that

$$\widetilde{\rho}_{p,r}(\widetilde{D}_\lambda^{-1}) := \inf \left\{ H \geq 0 : \widetilde{\rho}_r(\widetilde{D}_\lambda^{-1}(y)) \leq H \|y\|_p, y \in C(I, X) \right\} < +\infty.$$

First, we shall prove that if $p = 1$ and $r = \infty$, then $\widetilde{D}_\lambda^{-1}$ is bounded.

Theorem 51. *With the above notation,*

$$\widetilde{\rho}_{1,\infty}(\widetilde{D}_\lambda^{-1}) \leq 1.$$

Proof. We have to prove that

$$\widetilde{\rho}_\infty(\widetilde{D}_\lambda^{-1}(y)) \leq \|y\|_1, \qquad y \in C(I, X). \tag{2.8}$$

Note that, for each $y \in C(I, X)$ and $f \in C^1(I, X)$ with $D_\lambda f = y$ (i.e., $\widetilde{D}_\lambda^{-1}(y) = \widetilde{f}$), (2.8) takes the form

$$\widetilde{\rho}_\infty(\widetilde{f}) = \inf_{g \in N_\lambda} \|f - g\|_\infty \leq \|y\|_1 = \|D_\lambda f\|_1. \tag{2.9}$$

Clearly this is true if $\|y\|_1 = +\infty$. So, take $y \in C(I, X)$ with $\|y\|_1 < +\infty$. Then the equality $f' + \lambda f = y$ yields

$$f(t) = e^{-\lambda t} \left(\int_{t_0}^t e^{\lambda s} y(s) ds + k \right), \qquad t \in I,$$

for some $k \in X$. In order to prove (2.9) we need to distinguish two cases.

First, suppose that $\Re\lambda \geq 0$. Since $\|y\|_1 < +\infty$, it follows that

$$\int_a^{t_0} e^{\lambda s} y(s) ds \in X.$$

Let

$$g(t) := e^{-\lambda t}\left(k - \int_a^{t_0} e^{\lambda s} y(s) ds\right), \qquad t \in I.$$

Then $g \in N_\lambda$ and

$$\|f(t) - g(t)\| = \left\|e^{-\lambda t}\int_a^t e^{\lambda s} y(s) ds\right\|$$

$$\leq e^{-t\Re\lambda}\int_a^t e^{s\Re\lambda}\|y(s)\| ds \leq e^{-t\Re\lambda}\int_a^t e^{t\Re\lambda}\|y(s)\| ds$$

$$= \int_a^t \|y(s)\| ds \leq \int_a^b \|y(s)\| ds = \|y\|_1.$$

Therefore $\|f - g\|_\infty \leq \|y\|_1$, which means that (2.9) is proved.

Now consider the case when $\Re\lambda < 0$. Then

$$\int_{t_0}^b e^{\lambda s} y(s) ds \in X.$$

Let

$$g(t) := e^{-\lambda t}\left(k + \int_{t_0}^b e^{\lambda s} y(s) ds\right), \qquad t \in I.$$

Clearly, $g \in N_\lambda$ and

$$\|g(t) - f(t)\| = \left\|e^{-\lambda t}\int_t^b e^{\lambda s} y(s) ds\right\|$$

$$\leq e^{-t\Re\lambda}\int_t^b e^{s\Re\lambda}\|y(s)\| ds \leq e^{-t\Re\lambda}\int_t^b e^{t\Re\lambda}\|y(s)\| ds$$

$$= \int_t^b \|y(s)\| ds \leq \int_a^b \|y(s)\| ds = \|y\|_1, \qquad t \in I.$$

Again this implies (2.9). Thus the proof is finished. $\qquad\square$

Theorem 52. *If $b - a < +\infty$, $p \in [1, +\infty)$ and $r \in (0, +\infty)$, then*

$$\widetilde{\rho}_{p,r}(\widetilde{D}_\lambda^{-1}) \leq (b - a)^{1-1/p+1/r}. \tag{2.10}$$

Proof. Let $y \in C(I, X)$ and $f \in C^1(I, X)$ with $D_\lambda f = y$. As in the proof of Theorem 51, we construct a function $g \in N_\lambda$ such that

$$\|g(t) - f(t)\| \le \|y\|_1, \qquad t \in I. \tag{2.11}$$

According to Hölder's inequality,

$$\|y\|_1 \le (b-a)^{1-1/p} \|y\|_p,$$

so we get

$$\|g(t) - f(t)\| \le (b-a)^{1-1/p} \|y\|_p, \qquad t \in I.$$

Hence, it follows that

$$\|g - f\|_r = \left(\int_a^b \|g(t) - f(t)\|^r \, dt \right)^{1/r} \le (b-a)^{1-1/p+1/r} \|y\|_p.$$

Next, by (2.7), we have

$$\widetilde{\rho}_r(\widetilde{f}) \le (b-a)^{1-1/p+1/r} \|y\|_p.$$

On the other hand, $\widetilde{f} = \widetilde{D}_\lambda^{-1}(y)$, so that

$$\widetilde{\rho}_r(\widetilde{D}_\lambda^{-1}(y)) \le (b-a)^{1-1/p+1/r} \|y\|_p, \qquad y \in C(I, X).$$

Now (2.10) follows immediately (see Remark 7). $\qquad\qquad\qquad\qquad\qquad$ $\square$

Combining Theorems 51 and 52 we obtain the following:

Corollary 2. (i) *If $r = \infty$ and $p = 1$, then D_λ is Ulam stable with each constant $K > 1$.*

(ii) *If $b - a < +\infty$, $p \in [1, +\infty)$ and $r \in (0, +\infty)$, then D_λ is Ulam stable with each constant $K > (b-a)^{1-1/p+1/r}$.*

The results proved in Theorem 51 and Corollary 2 lead to the stability of the linear differential operator of order n with constant coefficients.

Now, let $a_1, a_2, \ldots, a_n \in \mathbb{K}$ and $D : C^n(I, X) \to C(I, X)$,

$$D(y) = y^{(n)} + a_1 y^{(n-1)} + \ldots + a_n y,$$

be a linear differential operator of order n. Denote by $\lambda_1, \lambda_2, \ldots, \lambda_n \in \mathbb{C}$ the roots of its characteristic equation, i.e.,

$$\lambda^n + a_1 \lambda^{n-1} + \ldots + a_n = 0.$$

Assume that $b - a < +\infty$, $\lambda_1, \lambda_2, \ldots, \lambda_n \in \mathbb{K}$ and the spaces $C^n(I, X)$ and $C(I, X)$ are endowed with the gauges $\|\cdot\|_r$ and $\|\cdot\|_p$, respectively. Then we have the following:

Theorem 53. (i) *If $r = \infty$ and $p = 1$, then D is Ulam stable with each constant*

$$K > (b - a)^{n-1}.$$

(ii) *If $p \in [1, \infty)$ and $r \in (0, +\infty)$, then D is Ulam stable with each constant*

$$K > (b - a)^{n-1/p+1/r}.$$

Proof. The operator D can be written as

$$D = D_{\lambda_1} \circ D_{\lambda_2} \circ \ldots \circ D_{\lambda_n}.$$

Now, the result follows from Corollary 2 and Theorem 37.

Let us only mention that, in the case of (i), the range of each D_{λ_k}, for $k = 1, \ldots, n$, is endowed with the gauge $\|\cdot\|_1$, the same the domain of D_{λ_k} for $k = 1, \ldots, n - 1$, and the domain of D_{λ_n} must be endowed with the gauge $\|\cdot\|_\infty$.

In the case of (ii) the range of D_{λ_k}, for $k = 1, \ldots, n$, is endowed with the gauge $\|\cdot\|_p$, while the domain of D_{λ_k} is endowed with the gauge $\|\cdot\|_r$ when $k = n$ and with the gauge $\|\cdot\|_p$ when $k < n$. $\square$

6. Some classical operators from the approximation theory

The goal of this section is to present some results on Ulam stability of some classical discrete and integral operators from the approximation theory (see [33]).

Let A and B be normed spaces and T a mapping from A into B. The following definition can be found in [40]:

Definition 6. We say that T has the *Ulam stability property* (briefly, T is *Ulam stable*) if there exists a constant K such that the following is true:
(i) For any $g \in T(A)$, $\varepsilon > 0$ and $f \in A$ with $\|Tf - g\| \le \varepsilon$, there exists an $f_0 \in A$ such that $Tf_0 = g$ and $\|f - f_0\| \le K\varepsilon$.

The number K is called an *Ulam constant of T*, and the infimum of all Ulam constants of T is denoted by K_T. Generally, K_T does not need to be an Ulam constant of T (see [16, 17]).

Now, let T be a bounded linear operator with kernel denoted by $N(T)$ and range denoted by $R(T)$. Consider the one-to-one operator $\widetilde{T}$ from the quotient space $A/N(T)$ into B:

$$\widetilde{T}(f + N(T)) = Tf, \qquad f \in A,$$

and the inverse operator $\widetilde{T}^{-1} : R(T) \to A/N(T)$.

In [40] the authors established the following result:

Theorem 54. ([40]) *Let A and B be Banach spaces and* $T : A \to B$ *be a bounded linear operator. Then the following statements are equivalent:*

(a) T *is Ulam stable;*
(b) $R(T)$ *is closed;*
(c) $\widetilde{T}^{-1}$ *is bounded.*

Moreover, if one of the conditions (a), (b), (c) is satisfied, then

$$K_T = \|\widetilde{T}^{-1}\|.$$

Remark 8. (1) Condition (i) in Definition 6 expresses the Ulam stability of the equation

$$Tf = g,$$

where $g \in R(T)$ is given and $f \in A$ is unknown.

(2) If $T : A \to B$ is a bounded linear operator, then (i) is equivalent to the following condition:

(ii) For any $f \in A$ with $\|Tf\| \leq 1$ there exists an $f_0 \in N(T)$ such that

$$\|f - f_0\| \leq K.$$

See also [17].

So, in what follows, we shall study the Ulam stability of a bounded linear operator $T : A \to B$ by checking the existence of a constant K for which (ii) is satisfied or, equivalently, by checking the boundedness of $\widetilde{T}^{-1}$.

Let $C[0, 1]$ be the space of all continuous, real-valued functions defined on $[0, 1]$, and $C_b[0, +\infty)$ the space of all continuous, bounded, real-valued functions on $[0, +\infty)$. Endowed with the supremum norm, they are Banach spaces.

6.1. Bernstein operators

For each integer $n \geq 1$ let Π_n be the subspace of $C[0, 1]$ consisting of all polynomial functions of degree $\leq n$. Let

$$p_{n,k}(x) := \binom{n}{k} x^k (1 - x)^{n-k}, \qquad x \in [0, 1], \ k \in \{0, 1, \ldots, n\}.$$

Then $\{p_{n,k}\}_{k=0,1,\ldots,n}$ is a basis of the linear space Π_n, called the *Bernstein-Bézier basis*.

The classical Bernstein operators $B_n : C[0, 1] \to C[0, 1]$ are defined by (see, e.g., [2])

$$B_n f = \sum_{k=0}^{n} p_{n,k} f\left(\frac{k}{n}\right), \qquad f \in C[0, 1], \ n \geq 1.$$

Let $n \geq 1$ be fixed. B_n is bounded, with $\|B_n\| = 1$. The range of the operator B_n is Π_n, and

$$N(B_n) = \left\{ f \in C[0,1] : f\left(\frac{k}{n}\right) = 0, \ k = 0, 1, \ldots, n \right\}.$$

Since $\dim \Pi_n = n + 1$, the operator $\widetilde{B}_n^{-1} : \Pi_n \to C[0,1]/N(B_n)$ is bounded. So, according to Theorem 41 and Remark 8.(1) we have

Theorem 55. *B_n is Ulam stable. Given $q \in \Pi_n$, the equation $B_n f = q$ (with unknown $f \in C[0,1]$) is stable in the Ulam sense.*

6.2. Szász-Mirakjan operators

The nth Szász-Mirakjan operator $L_n : C_b[0, +\infty) \to C_b[0, +\infty)$ is defined by (see, e.g., [2, p. 338])

$$L_n f(x) = e^{-nx} \sum_{i=0}^{\infty} f\left(\frac{i}{n}\right) \frac{n^i}{i!} x^i, \qquad x \in [0, +\infty).$$

Theorem 56. *For each $n \geq 1$ the operator L_n is not Ulam stable.*

Proof. Suppose that for a certain $n \geq 1$, L_n is Ulam stable. Then there exists a constant K such that for any $f \in C_b[0, +\infty)$ with $\|L_n f\|_\infty \leq 1$ there exists $g \in N(L_n)$ with

$$\|f - g\|_\infty \leq K.$$

According to Stirling's formula,

$$\lim_{i \to \infty} \frac{i^i}{i! e^i} = 0,$$

so that there exists a $j \geq 1$ such that

$$(K + 1)\frac{j^j}{j! e^j} \leq 1.$$

Let $f \in C_b[0, +\infty)$ be the function defined by

$$f(x) = 0, \qquad x \in \left[0, \frac{j-1}{n}\right] \cup \left[\frac{j+1}{n}, +\infty\right);$$

$$f\left(\frac{j}{n}\right) = K + 1;$$

and f be linear on $\left[\frac{j-1}{n}, \frac{j}{n}\right]$ and on $\left[\frac{j}{n}, \frac{j+1}{n}\right]$. Then

$$L_n f(x) = e^{-nx} \cdot \frac{n^j}{j!} x^j (K+1), \qquad x \in [0, +\infty).$$

It is easy to check that $\|L_n f\|_\infty \leq 1$, so that there exists $g \in N(L_n)$ with $\|f - g\|_\infty \leq K$. But then $g\left(\frac{j}{n}\right) = 0$, and consequently

$$K \geq \|f - g\|_\infty \geq \left| f\left(\frac{j}{n}\right) - g\left(\frac{j}{n}\right) \right| = K + 1,$$

which is a contradiction. Thus the theorem is proved. $\qquad\qquad\qquad\square$

6.3. Other classical operators

The method of proof used in the case of Szász-Mirakjan operators can be also used in order to prove that the Meyer-König and Zeller operators $L_n : C[0, 1] \to C[0, 1]$

$$L_n f(x) = (1 - x)^{n+1} \sum_{i=0}^{\infty} f\left(\frac{i}{n+i}\right) \binom{n+i}{i} x^i, \qquad x \in [0, 1),$$

$$L_n f(1) = f(1),$$

as well as the Baskakov operators $L_n : C_b[0, +\infty) \to C_b[0, +\infty)$

$$L_n f(x) = \sum_{i=0}^{\infty} f\left(\frac{i}{n}\right) \binom{n+i-1}{i} \frac{x^i}{(1+x)^{n+i}}, \qquad x \in [0, +\infty),$$

are not Ulam stable.

On the other hand the Stancu operators [2, p. 301], the Bleimann-Butzer-Hahn operators [2, p. 306], the Bernstein-Schurer operators [2, p. 320], the Bernstein-Cheney-Sharma operators [2, p. 322], the Fejér-Hermite operators [2, p. 331], and the Bernstein operators of second kind [37] are discrete Ulam stable operators since their ranges are finite-dimensional.

6.4. Integral operators

Consider the Kantorovich operators ([2, p. 333]) $K_n : C[0, 1] \to C[0, 1]$,

$$K_n f(x) := (n+1) \sum_{k=0}^{n} p_{n,k}(x) \int_{\frac{k}{n+1}}^{\frac{k+1}{n+1}} f(t) dt, \qquad x \in [0, 1],$$

the Durrmeyer operators ([2, p. 335]) $M_n : C[0, 1] \to C[0, 1]$,

$$M_n f(x) := (n + 1) \sum_{k=0}^{n} p_{n,k}(x) \int_0^1 p_{n,k}(t) f(t) dt, \qquad x \in [0, 1],$$

and the genuine Bernstein-Durrmeyer operators (see, e.g., [15] and the references therein) $U_n : C[0, 1] \to C[0, 1]$,

$$U_n f(x) := f(0) p_{n,0}(x) + f(1) p_{n,n}(x) + (n - 1) \sum_{k=1}^{n-1} p_{n,k}(x) \int_0^1 p_{n-2,k-1}(t) f(t) dt$$

for $x \in [0, 1]$.

They are Ulam stable, since their ranges are finite-dimensional.

Now let us consider the Beta operator introduced by A. Lupaş [26]:

$$L_n f(x) := \frac{\displaystyle\int_0^1 t^{nx} (1 - t)^{n(1-x)} f(t) dt}{\displaystyle\int_0^1 t^{nx} (1 - t)^{n(1-x)} dt},$$

where $n \geq 1$, $f \in C[0, 1]$, $x \in [0, 1]$.

Theorem 57. *For each $n \geq 1$, the Beta operator $L_n : C[0, 1] \to C[0, 1]$ is not Ulam stable.*

Proof. The operator L_n is injective.

Indeed, let $n \geq 1$ be a given integer and suppose that $L_n f = 0$ for a certain $f \in C[0, 1]$. Then

$$\int_0^1 t^{nx} (1 - t)^{n(1-x)} f(t) dt = 0, \qquad x \in [0, 1].$$

The change of variables,

$$\frac{t}{1 - t} = u,$$

leads to

$$\int_0^\infty u^{nx} \frac{1}{(1 + u)^{n+2}} f\left(\frac{u}{1 + u}\right) du = 0.$$

Let g be the continuous function defined by

$$g(u) := \frac{1}{(1 + u)^{n+2}} f\left(\frac{u}{1 + u}\right), \qquad u \in [0, \infty).$$

The equality

$$\int_0^\infty u^{nx} g(u)\,du = 0, \qquad x \in [0, 1]$$

can be rewritten as

$$M[g](nx + 1) = 0, \qquad x \in [0, 1],$$

where $M[g]$ denotes the Mellin Transform of g (see [9, 12]). Put $nx + 1 = s$, $s \in [1, n + 1]$. The equality

$$M[g](s) = M[0](s), \qquad s \in [1, n + 1],$$

leads to $g(u) = 0$ a.e. on $[0, \infty)$, according to [12, Theorem 1.5.22, p. 57]. Now taking account of the continuity of g it follows $g(u) = 0$ for all $u \in [0, \infty)$ and finally $f(t) = 0$, for all $t \in [0, 1]$. Therefore L_n is injective. (For another proof of the injectivity of L_n see [14]).

Let us consider the inverse operator $L_n^{-1} : R(L_n) \to C[0, 1]$.

Denote $e_j(x) = x^j$, $j = 0, 1, 2, \ldots$; $x \in [0, 1]$. It is easy to verify that $L_n e_0 = e_0$, and

$$L_n e_j(x) = \frac{(nx + 1)(nx + 2)\ldots(nx + j)}{(n + 2)(n + 3)\ldots(n + j + 1)}, \qquad j = 1, 2, \ldots.$$

It follows that for each $j \geq 1$,

$$\frac{n^j}{(n + 2)\ldots(n + j + 1)}$$

is an eigenvalue of L_n, and consequently

$$\frac{(n + 2)\ldots(n + j + 1)}{n^j}$$

is an eigenvalue of L_n^{-1}.

Since

$$\lim_{j \to \infty} \frac{(n + 2)\ldots(n + j + 1)}{n^j} = +\infty$$

we conclude that L_n^{-1} is unbounded, so L_n is not Ulam stable.   □

6.5. Bernstein-Schnabl operators

Ulam stability of Bernstein-Schnabl operators is closely related to the Fréchet functional equation. We present some results obtained in [32].

Let $f : \mathbb{R} \to \mathbb{R}$ and $n \in \mathbb{N}$. Consider the difference operator of order n, defined as usual by

$$\Delta_h^n f(x) := \sum_{j=0}^{n} (-1)^j \binom{n}{j} f(x + (n-j)h), \tag{2.12}$$

where $x \in \mathbb{R}$ and $h \in \mathbb{R}$ (see [10, p. 504]).

If f is a polynomial function of degree at most $n - 1$, then it is well known that

$$\Delta_h^n f(x) = 0, \tag{2.13}$$

for all $x \in \mathbb{R}$ and $h \in \mathbb{R}$ (see [1] and [24, p. 271]).

One may ask if the converse is also true. If $n > 1$, without any regularity assumptions on f the answer is negative.

In 1909 M. Fréchet [11] studied an equation more general than (2.13), which characterizes the polynomials among the continuous mappings. From the result of Fréchet it follows that a continuous function $f : \mathbb{R} \to \mathbb{R}$ satisfies (2.13) for all $x \in \mathbb{R}$ and $h \in \mathbb{R}$ if and only if f is a polynomial function of the degree, at most, $n - 1$; for details see [1, Theorem 1] and also [24, Theorem 13.5]. Due to the result of Fréchet, equation (2.13) bears his name.

Th. Angheluţă [5, 6] (see also [24]) proved that the only measurable functions satisfying equation (2.13) for all $h > 0$ and all $x \in \mathbb{R}$ are polynomials of degree at most $n - 1$. A stronger result concerning polynomial solutions of (2.13) was obtained by P. Montel [28, 29] and T. Popoviciu [35] (see also [24]) under different regularity assumptions on f but weaker on h. Recently J.M. Almira and A.J. López-Moreno [1] gave a new proof of the classical result of Fréchet, assuming about f only the continuity at some point or the boundedness on some nonempty open set.

All the above results concerning the Fréchet functional equation are obtained under the assumption that (2.13) is satisfied for all values of x and some values of h. In [32] it is proved that similar results hold assuming that (2.13) is satisfied for all values of h in a given set and a unique value of x.

This approach is motivated by the study of the Ulam stability of some linear operators.

Let $n \geq 1$ be fixed. We are looking for functions $f \in C^n[0, 1]$ that satisfy the equation

$$\Delta_t^n f(0) = 0 \text{ for all } t \in \left[0, \frac{1}{n}\right]. \tag{2.14}$$

The nth order difference operator Δ_t^n is defined as in (2.12) so that the Fréchet-type functional equation (2.14) can be written as

$$\sum_{j=0}^{n} (-1)^j \binom{n}{j} f((n-j)t) = 0, \qquad t \in \left[0, \frac{1}{n}\right]. \tag{2.15}$$

Taking into account the definition of Let $k(n)$ be defined as follows:

$$k(1) := 0, \qquad k(2) := 1, \qquad k(3) := 3,$$

$$k(n) := \min\left\{k \in \mathbb{N} : \sum_{i=1}^{n-1} \binom{n}{i}\left(1 - \frac{i}{n}\right)^k < 1\right\}, \qquad n \geq 4.$$

Let $n \geq 4$ and $k \leq n$. Then

$$\sum_{i=1}^{n-1} \binom{n}{i}\left(1 - \frac{i}{n}\right)^k > \binom{n}{1}\left(1 - \frac{1}{n}\right)^k \geq n\left(1 - \frac{1}{n}\right)^n > 1.$$

Thus $k(n) > n$ for all $n \geq 4$. In particular, $k(4) = 6$.

Let Π_m be the subspace of $C[0, 1]$ consisting of all polynomial functions of degree at most m. Each $f \in \Pi_{n-1}$ is a solution of the Fréchet equation (2.13), hence a solution of (2.14). The next result goes in the opposite direction.

Theorem 58. *If $f \in C^{k(n)}[0, 1]$ is a solution of the functional equation* (2.14), *then $f \in \Pi_{n-1}$.*

Proof. The assertion is trivial for $n = 1$. Let $n = 2$ and suppose that $f \in C^1[0, 1]$ is a solution to

$$\Delta_t^2 f(0) := f(2t) - 2f(t) + f(0) = 0, \qquad t \in \left[0, \frac{1}{2}\right].$$

Then $f'(2t) = f'(t)$, $t \in \left[0, \frac{1}{2}\right]$, and it is easy to conclude that $f'(x) = f'(2^{-j}x)$ for all $x \in [0, 1]$ and $j \in \mathbb{N}$. It follows that $f'(x) = f'(0)$, $x \in [0, 1]$, which entails $f \in \Pi_1$.

Let $n = 3$ and $f \in C^3[0, 1]$ such that

$$\Delta_t^3 f(0) := f(3t) - 3f(2t) + 3f(t) - f(0) = 0, \qquad t \in \left[0, \frac{1}{3}\right]. \tag{2.16}$$

Setting $\varphi := f^{(3)}$ we get

$$\varphi(x) = \frac{8}{9}\varphi\left(\frac{2}{3}x\right) - \frac{1}{9}\varphi\left(\frac{1}{3}x\right), \qquad x \in [0, 1]. \tag{2.17}$$

Let $s \in [0, 1]$ with $|\varphi(s)| = \|\varphi\|_\infty$. Then

$$\|\varphi\|_\infty = |\varphi(s)| \leq \frac{8}{9}\left|\varphi\left(\frac{2}{3}s\right)\right| + \frac{1}{9}\left|\varphi\left(\frac{1}{3}s\right)\right| \leq \|\varphi\|_\infty.$$

This yields

$$\left|\varphi\left(\frac{1}{3}s\right)\right| = \|\varphi\|_\infty,$$

i.e.,

$$\|\varphi\|_\infty = \left|\varphi\left(\frac{1}{3^j}s\right)\right|, \qquad j \geq 1.$$

Since $\varphi \in C[0, 1]$, we get $\|\varphi\|_\infty = |\varphi(0)|$. From (2.17) we see that $\varphi(0) = 0$, so that $\varphi = 0$; this entails $f^{(3)} = 0$, hence $f \in \Pi_2$.

It remains to consider the case $n \geq 4$.

Suppose that $f \in C^{k(n)}[0, 1]$ satisfies (2.14). Setting $\varphi := f^{(k(n))}$ we obtain easily

$$\sum_{j=0}^{n-1} (-1)^j \binom{n}{j}(n-j)^{k(n)}\varphi((n-j)t) = 0, \qquad t \in \left[0, \frac{1}{n}\right].$$

Denoting $t = \frac{x}{n}$ we infer that

$$\varphi(x) = \sum_{j=1}^{n-1} (-1)^{j-1}\binom{n}{j}\left(1-\frac{j}{n}\right)^{k(n)}\varphi\left(\left(1-\frac{j}{n}\right)x\right), \qquad x \in [0, 1].$$

Consequently,

$$\|\varphi\|_\infty \leq \|\varphi\|_\infty \sum_{j=1}^{n-1} \binom{n}{j}\left(1-\frac{j}{n}\right)^{k(n)}.$$

Taking into account the definition of $k(n)$ we conclude that $\|\varphi\|_\infty = 0$, i.e., $f \in \Pi_{k(n)-1}$.

Since $f \in C^n[0, 1]$, for each $0 < x \leq \frac{1}{n}$ there exists $u(x) \in (0, nx)$ such that

$$0 = \Delta_x^n f(0) = x^n f^{(n)}(u(x)).$$

(See [10, p. 505] and [39, Theorem 2.10]).

Thus, for each $0 < x \leq \frac{1}{n}$ there exists $u(x) \in (0, nx)$ such that $f^{(n)}(u(x)) = 0$.

Consider the sequence $(x_j)_{j\geq 1}$ defined as follows:

$$x_1 := \frac{1}{n}, \qquad x_{j+1} := \frac{1}{n}u(x_j), \qquad j \geq 1.$$

Then, for all $j \geq 1$,

$$0 < u(x_{j+1}) < nx_{j+1} = u(x_j) < nx_j \leq 1.$$

We see that the polynomial function $f^{(n)} \in \Pi_{k(n)-n-1}$ has infinitely many roots, namely $u(x_j)$, $j \geq 1$. This means that $f^{(n)} = 0$, i.e. $f \in \Pi_{n-1}$. $\qquad\square$

Now we are in position to prove Ulam stability of Bernstein-Schanbl operators.

Consider a continuous selection of probability Borel measures on $[0, 1]$, i.e., a family $(\mu_x)_{0 \leq x \leq 1}$ of probability Borel measures on $[0, 1]$ such that for every $f \in C[0, 1]$ the function

$$x \mapsto \int_0^1 f d\mu_x$$

is continuous on $[0, 1]$. Suppose that

$$\int_0^1 t d\mu_x(t) = x, \qquad x \in [0, 1].$$

For every $n \geq 1$, the n-th Bernstein-Schnabl operator associated with the selection $(\mu_x)_{0 \leq x \leq 1}$ is the positive linear operator $B_n : C[0, 1] \to C[0, 1]$ defined for every $f \in C[0, 1]$ and $x \in [0, 1]$ as

$$B_n f(x) := \int_0^1 \cdots \int_0^1 f\left(\frac{x_1 + \ldots + x_n}{n}\right) d\mu_x(x_1) \ldots d\mu_x(x_n).$$

This definition can be found in [4], where the properties of the sequence $(B_n)_{n \geq 1}$ and of the associated Markov semigroup are deeply investigated; see also [36, 38].

Here we consider a special selection and investigate the associated operators B_n from the point of view of the Ulam stability. Theorem 58 will be a tool in this investigation.

Let $u \in C[0, 1]$, $u(x) = \min\{x, 1 - x\}$, $x \in [0, 1]$. Consider the family $(\mu_x)_{0 \leq x \leq 1}$ of probability Borel measures on $[0, 1]$ such that for every $f \in C[0, 1]$,

$$\int_0^1 f d\mu_x := \begin{cases} f(x), & x \in \{0, 1\}, \\ \dfrac{1}{2u(x)} \displaystyle\int_{x-u(x)}^{x+u(x)} f(t) dt, & 0 < x < 1. \end{cases}$$

The associated Bernstein-Schnabl operators are in this case defined by

$$B_n f(x) := \begin{cases} f(x), & x \in \{0, 1\}, \\ \beta(x), & 0 < x < 1, \end{cases} \tag{2.18}$$

for all $n \geq 1$, $f \in C[0, 1]$, $x \in [0, 1]$, where

$$\beta(x) := \frac{1}{(2u(x))^n} \int_{x-u(x)}^{x+u(x)} \cdots \int_{x-u(x)}^{x+u(x)} f\left(\frac{x_1 + \ldots + x_n}{n}\right) dx_1 \ldots dx_n.$$

Let $L : X \to Y$ be a bounded linear operator acting between the normed spaces X

and Y. One way of expressing the Ulam stability of L is the following (see [16, 17, 40]):

L is stable in Ulam sense if there exists a real number $K > 0$ such that for each $x \in X$ with $\|Lx\| \leq 1$ there is an $x' \in X$ satisfying $Lx' = 0$ and $\|x' - x\| \leq K$.

This is equivalent to saying that the equation $Lx = y$ is Ulam stable, i.e.,

For each $\varepsilon > 0$, $y \in L(X)$ and $z \in X$ with $\|Lz - y\| \leq \varepsilon$ there is a $z' \in X$ satisfying $Lz' = y$ and $\|z' - z\| \leq K\varepsilon$.

The next result is concerned with the restriction of operator B_n to the normed space $(X_n, \| \cdot \|_\infty)$, where

$$X_n = \begin{cases} C[0,1], & n \in \{1,2,3\}, \\ C^{k(n)-n}[0,1], & n \geq 4. \end{cases}$$

Theorem 59. *For each $n \geq 1$, $B_n : X_n \to C[0,1]$ is not stable in the sense of Ulam.*

Proof. Let $n \geq 1$ be fixed. Let $f \in X_n$ with $B_n f = 0$. Set

$$m(n) := \max\{n, k(n)\}$$

and let $F \in C^{m(n)}[0,1]$ with $F^{(n)} = f$.

Let $0 < x \leq \dfrac{1}{2}$. According to (2.18),

$$B_n f(x) = \frac{1}{(2x)^n} \int_0^{2x} \cdots \int_0^{2x} f\left(\frac{x_1 + \ldots + x_n}{n}\right) dx_1 \ldots dx_n.$$

Using the integral representation of a finite difference (see [39, Theorem 2.9]) we get

$$B_n f(x) = \frac{n^n}{(2x)^n} \Delta_{\frac{2x}{n}}^n F(0).$$

Since $B_n f(x) = 0$, we obtain

$$\Delta_{\frac{2x}{n}}^n F(0) = 0, \qquad 0 < x \leq \frac{1}{2}. \tag{2.19}$$

Obviously (2.19) is valid also for $x = 0$. Denoting $t := \dfrac{2x}{n}$, we find

$$\Delta_t^n F(0) = 0, \qquad t \in \left[0, \frac{1}{n}\right].$$

Now Theorem 58 shows that $F \in \Pi_{n-1}$, i.e., $f = F^{(n)} = 0$.

Suppose that there exists a number $K > 0$ such that for all $g \in X_n$ with $\|B_n g\|_\infty \leq 1$ there is an $f \in X_n$ with $B_n f = 0$ and $\|g - f\|_\infty \leq K$. Such an f is necessarily the null function, so for each $g \in X_n$ with $\|B_n g\|_\infty \leq 1$ we have $\|g\|_\infty \leq K$. Let us show that this leads to a contradiction.

For each $0 < a < \dfrac{1}{2}$ take a function $g_a \in X_n$ such that

$$0 \leq g_a(x) \leq g_a\left(\frac{1}{2}\right) = K + 1, \qquad x \in [0, 1],$$

and $g_a(x) = 0$ for $x \in [0, a) \cup (1 - a, 1]$.

Then $\|g_a\|_\infty = K + 1$, and it remains only to show that

$$\lim_{a \to \frac{1}{2}} \|B_n g_a\|_\infty = 0.$$

For

$$0 < x \leq \frac{1}{2}$$

we have

$$B_n g_a(x) = \frac{1}{(2x)^n} \int_0^{2x} \cdots \int_0^{2x} g_a\left(\frac{x_1 + \ldots + x_n}{n}\right) dx_1 \ldots dx_n$$

$$\leq \frac{K + 1}{(2x)^n} \mathrm{Vol}_n \Delta(x, a),$$

where

$$\Delta(x, a) := \{(x_1, \ldots, x_n) \in [0, 2x]^n : na \leq x_1 + \ldots + x_n \leq n(1 - a)\}$$

and Vol_n stands for the n-dimensional volume.

For

$$0 < x < \frac{a}{2},$$

$\Delta(x, a)$ is empty, and so $B_n g_a(x) = 0$.

For

$$\frac{a}{2} \leq x \leq \frac{1}{2},$$

we have

$$\mathrm{Vol}_n \Delta(x, a) \leq \mathrm{Vol}_n \Delta\left(\frac{1}{2}, a\right),$$

and so

$$B_n g_a(x) \leq \frac{K + 1}{(2x)^n} \mathrm{Vol}_n \Delta\left(\frac{1}{2}, a\right) \leq \frac{K + 1}{a^n} \mathrm{Vol}_n \Delta\left(\frac{1}{2}, a\right).$$

Since $B_n g_a(0) = g_a(0) = 0$, we conclude that

$$0 \le B_n g_a(x) \le \frac{K+1}{a^n} Vol_n \Delta \left(\frac{1}{2}, a \right)$$

for all $x \in \left[0, \frac{1}{2} \right]$.

Now consider the function $\widetilde{g}_a(t) = g_a(1 - t)$, $t \in [0, 1]$. It is not difficult to verify that for $x \in \left[\frac{1}{2}, 1 \right]$ one has

$$0 \le B_n g_a(x) = B_n \widetilde{g}_a(1 - x) \le \frac{K+1}{a^n} Vol_n \Delta \left(\frac{1}{2}, a \right).$$

Consequently,

$$\|B_n g_a\| \le \frac{K+1}{a^n} Vol_n \Delta \left(\frac{1}{2}, a \right), \qquad 0 < a < \frac{1}{2}.$$

Since

$$\lim_{a \to \frac{1}{2}} Vol_n \Delta \left(\frac{1}{2}, a \right) = 0,$$

we see that

$$\lim_{a \to \frac{1}{2}} \|B_n g_a\|_\infty = 0,$$

and this concludes the proof. $\square$

Remark 9. The regularity of f, required in Theorem 58, is important. For example, let $n = 2$; we shall construct nonpolynomial functions $f \in C[0, 1]$ with

$$\Delta_t^2 f(0) = 0, \qquad t \in \left[0, \frac{1}{2} \right].$$

Indeed, let f be a nonpolynomial continuous function on $\left[\frac{1}{2}, 1 \right]$, with

$$f \left(\frac{1}{2} \right) = \frac{1}{2}$$

and $f(1) = 1$. For each $j \in \{1, 2, \ldots\}$ and $t \in [2^{-j-1}, 2^{-j})$ set $f(t) := 2^{-j} f(2^j t)$. Finally set $f(0) := 0$. Then $f \in C[0, 1]$ and

$$f(t) = \frac{1}{2} f(2t), \qquad t \in \left[0, \frac{1}{2} \right],$$

which entails $f(2t) - 2f(t) + f(0) = 0$, i.e.,

$$\Delta_t^2 f(0) = 0, \qquad t \in \left[0, \frac{1}{2}\right].$$

Open problem. It would be interesting to know if for $n \geq 4$ Theorem 58 holds with $f \in C^n[0, 1]$ instead of $f \in C^{k(n)}[0, 1]$; in this case, Theorem 59 would be valid with $C[0, 1]$ instead of X_n.

REFERENCES

1. J.M. Almira, A.J. López-Moreno, On solutions of the Fréchet functional equation, J. Math. Anal. Appl. 332 (2007) 1119–1133.
2. F. Altomare, M. Campiti, Korovkin-type Approximation Theory and its Applications, W. de Gruyter, Berlin - New York, 1994.
3. F. Altomare, M. Cappelletti Montano, V. Leonessa, I. Raşa, Markov Operators, Positive Semigroups and Approximation Processes, De Gruyter Studies in Mathematics, Vol. 61, 2014.
4. F. Altomare, V. Leonessa, I. Raşa, On Bernstein-Schnabl operators on the unit interval, Zeit. Anal. Anwend. 27 (2008), 353–379.
5. Th. Angheluţă, Sur une équation fonctionnelle caractérisant les polynômes, Mathematica (Cluj) 6 (1932) 1–7.
6. Th. Angheluţă, Sur une propriété des polynômes, Bull. Sci. Math. Paris 63 (1939) 239–246.
7. J. Brzdęk, S.M. Jung, A note on stability of an operator linear equation of the second order, Abstr. Appl. Anal. 2011 (2011), Art. ID 602713, 15 pp.
8. J. Brzdęk, D. Popa, I. Raşa, Hyers-Ulam stability with respect to gauges, J. Math. Anal. Appl. 453 (2017) 620–628.
9. P.L. Butzer, S. Jansche, A direct approach to the Mellin Transform, J. Fourier Anal. Appl. 3 (1997) 325–376.
10. B. Demidovitch, I. Maron, Éléments de Calcul Numérique, Edition MIR, Moscou, 1973.
11. M. Fréchet, Une définition fonctionnelle des polynomes, Nouv. Ann. Math. 9 (1909) 145–162.
12. H.J. Glaeske, A.P. Prudnikov, K.A. Skòrnik, Operational Calculus and Related Topics, Chapman & Hall/CRC, Taylor & Francis Group, 2006.
13. J. A. Goldstein, Semigroups of Linear Operators and Applications, Oxford Univ. Press, New York, 1985.
14. H. Gonska, M. Heilmann, A. Lupaş, I. Raşa, On the composition and decomposition of positive linear operators III: a non-trivial decomposition of the Bernstein operator, Univ. Duisburg - Essen, Technical Report SM-DU-739, 2011.
15. H. Gonska, D. Kacsó, I. Raşa, On genuine Bernstein-Durrmeyer operators, Result. Math. 50 (2007) 213–225.
16. O. Hatori, K. Kobayashi, T. Miura, H. Takagi, S.E. Takahasi, On the best constant of Hyers-Ulam stability, J. Nonlinear Convex Anal. 5 (2004) 387–393.
17. G. Hirasawa, T. Miura, Hyers-Ulam stability of a closed operator in a Hilbert space, Bull. Korean. Math. Soc. 43 (2006) 107–117.
18. Q.L. Huang, M.S. Moslehian, Relationship between the Hyers-Ulam stability and the Moore-Penrose inverse, Electron. J. Linear Algebra 23 (2012) 891-905.
19. Q.L. Huang, L. Zhu, B. Wu, A Note on the Hyers-Ulam stability constants of closed linear operators, Filomat 29 (2015) 909–915.
20. S. Karlin, W.J. Studden, Tchebycheff Systems: with Applications in Analysis and Statistics, Inter-science Publishers, 1966.
21. T. Kato, Perturbation Theory for Linear Operators, Springer, Berlin 1995.
22. U. Krengel, M. Lin, On the range of the generator of a Markovian semigroup, Math. Z. 185 (1984) 553–565.

23. E. Kreyszig, Introductory Functional Analysis with Applications, Wiley, 1978.
24. M. Kuczma, Functional Equations in a Single Variable, Polish Scientific Publishers, Warszawa, 1968.
25. M. Lin, On the uniform ergodic theorem. II, Proc. Amer. Math. Soc. 46 (1974) 217–225.
26. A. Lupaş, Die Folge der Betaoperatoren, Dissertation, Univ. Stuttgart, 1972.
27. T. Miura, S. Miyajima, S.E. Takahasi, Hyers-Ulam stability of linear differential operator with constant coefficients, Math. Nachr. 258 (2003) 90–96.
28. P. Montel, Sur un théorème de Jacobi, C. R. Acad. Sci. Paris 201 (1935) 586–588.
29. P. Montel, Sur des équations fonctionnelles caracterisant les polynomes, C. R. Acad. Sci. Paris 226 (1948) 1053–1055.
30. M.S. Moslehian, G. Sadeghi, Perturbation of closed range operators, Turk. J. Math. 33 (2009) 143–149.
31. A. Pazy, Semigroups of Linear Operators and Applications to Partial Differential Equations, Springer, New-York, 1983.
32. D. Popa, I. Raşa, The Fréchet functional equation with application to the stability of certain operators, J. Approx. Theory, 164 (2012) 138–144.
33. D. Popa, I. Raşa, On the stability of some classical operators from approximation theory, Expo. Math. 31 (2013) 205–214.
34. D. Popa, I. Raşa, Best constant in stability of some positive linear operators, Aequationes Math. 90 (2016) 719–726.
35. T. Popoviciu, Remarques sur la définition fonctionnelle d'un polynôme d'une variable réelle, Mathematica (Cluj) 12 (1936) 5–12.
36. I. Raşa, Asymptotic behaviour of iterates of positive linear operators, Jaén J. Approx. 1 (2009) 195–204.
37. I. Raşa, Classes of convex functions associated with Bernstein operators of second kind, Math. Inequal. Appl. 9 (2006) 599–605.
38. I. Raşa, C_0-semigroups and iterates of positive linear operators: asymptotic behaviour, Rend. Circ. Mat. Palermo, Ser. II, Suppl. 82 (2010) 123–142.
39. P.K. Sahoo, T. Riedel, Mean Value Theorems and Functional Equations, World Scientific, 1998.
40. H. Takagi, T. Miura, S.E. Takahasi, Essential norms and stability constants of weighted composition operators on $C(X)$, Bull. Korean Math. Soc. 40 (2003) 583–591.
41. T. Vladislav, I. Raşa, Analiză numerică. Aproximare, problema lui Cauchy abstractă, proiectori Altomare, Editura Tehnică, Bucureşti, 1999.

CHAPTER 3

Ulam stability of differential operators

Contents

Abstract

The chapter contains results on the Ulam stability for linear differential equations, linear differential operators, and partial differential equations in Banach spaces. As a consequence we improve some known estimates of the difference between the perturbed and the exact solution.

1. Introduction

M. Obłoza seems to be the first author to investigate Ulam's stability of differential equations [20, 21]. Later C. Alsina and R. Ger proved that for every differentiable mapping $f : I \to \mathbb{R}$ satisfying

$$|f'(x) - f(x)| \leq \varepsilon, \qquad x \in I,$$

where $\varepsilon > 0$ is a given number and I is an open interval of $\mathbb{R}$, there exists a differentiable function $g : I \to \mathbb{R}$ with the property

$$g'(x) = g(x), \qquad |f(x) - g(x)| \leq 3\varepsilon, \qquad x \in I.$$

The result of Alsina and Ger [2] was extended by T. Miura, S. Miyajima, and S.E. Takahasi [18, 19, 31] and by S.E. Takahasi, H. Takagi, T. Miura, and S. Miyajima [32] to the Ulam stability of the first-order linear differential equations and linear differential equations of higher order with constant coefficients. Furthermore S.M. Jung

http://dx.doi.org/10.1016/B978-0-12-809829-5.50003-9
Copyright © 2018 Elsevier Inc. All rights reserved.

[11, 12, 13, 14] obtained results on the stability of linear differential equations extending the results of Takahasi, Takagi, and Miura. I.A. Rus obtained some results on the stability of differential and integral equations using the Gronwall lemma and the technique of weakly Picard operators [28, 29]. G. Wang, M. Zhou, L. Sun [33], and Y. Li, Y. Shen [15] proved the Ulam stability of the linear differential equation of the first order and the linear differential equation of the second order with constant coefficients by using the integral factor method.

An extension of the results given in [12, 15, 19] was obtained by D.S. Cîmpean, D. Popa [6] and D. Popa, I. Raşa [23] for the linear differential equation of the n-th order with constant coefficients and the linear differential operator of the n-th order with nonconstant coefficients. It seems that the first paper on Ulam stability of partial differential equations was written by A. Prástaro and Th.M. Rassias [24]. For recent results on this subject we refer the reader to [7, 16, 17, 27, 30]

Throughout this chapter by $(X, \| \cdot \|)$ we denote a Banach space over the field $\mathbb{K}$ ($\mathbb{K}$ is one of the fields $\mathbb{R}$ or $\mathbb{C}$), unless explicitly stated otherwise. In what follows by $\mathfrak{R}z$ we denote the real part of the complex number z.

Recall first some results obtained on Ulam stability of the first order linear differential operator. Let X be a complex Banach space, $h : \mathbb{R} \to \mathbb{C}$ be a continuous function, and consider the linear operator $T_h : C^1(\mathbb{R}, X) \to C(\mathbb{R}, X)$ given by

$$T_h u = u' + hu, \qquad u \in C^1(\mathbb{R}, X).$$

Define

$$\|u\|_\infty = \sup_{t \in \mathbb{R}} \|u(t)\|$$

for some function $u : \mathbb{R} \to X$. Clearly $\|\cdot\|_\infty$ is a gauge.

The operator T_h is Ulam stable if there exists $K \geq 0$ such that for every $\varepsilon > 0$, $u \in C^1(\mathbb{R}, X)$, $v \in C(\mathbb{R}, X)$ such that

$$\|T_h u - v\|_\infty \leq \varepsilon$$

there exists $u_0 \in C^1(\mathbb{R}, X)$ with the properties $T_h u_0 = v$ and

$$\|u - u_0\|_\infty \leq K \cdot \varepsilon.$$

We call such K an Ulam constant for T_h. If, in addition, minimum of such K's exists, then we call it the best Ulam constant. For a continuous function $h : \mathbb{R} \to \mathbb{C}$ we define

$$\widetilde{h}(t) = \exp\left(\int_0^t h(s)ds \right), \quad t \in \mathbb{R},$$

and let

$$C_h = \sup_{t \in \mathbb{R}} \frac{1}{\left|\widetilde{h}(t)\right|} \int_t^\infty \left|\widetilde{h}(s)\right| ds,$$

$$D_h = \sup_{t \in \mathbb{R}} \frac{1}{\left|\widetilde{h}(t)\right|} \int_{-\infty}^t \left|\widetilde{h}(s)\right| ds,$$

$$E_h = \sup_{t \in \mathbb{R}} \frac{1}{\left|\widetilde{h}(t)\right|} \left|\int_0^t \left|\widetilde{h}(s)\right| ds\right|.$$

A characterization of Ulam stability of T_h is given in the next theorem.

Theorem 60 ([18]). *Let $h : \mathbb{R} \to \mathbb{C}$ be a continuous function. Then T_h has Ulam stability if and only if one of C_h, D_h, E_h is finite. Moreover, the following statements are valid:*

(i)   *If C_h is finite, then C_h is an Ulam constant for T_h;*
(ii) *If D_h is finite, then D_h is an Ulam constant for T_h;*
(iii) *If E_h is finite, then E_h is an Ulam constant for T_h.*

The operator T_h is not Ulam stable if and only if

$$C_h = D_h = E_h = +\infty.$$

As a consequence it follows that if $h(t) = \lambda$, for all $t \in \mathbb{R}$ and $\lambda \in \mathbb{C}$, then T_h is Ulam stable if and only if $\mathfrak{R}\lambda \neq 0$.

The previous outcome leads to a result on Ulam stability for the linear differential operator of order n with constant coefficients, i.e.,

$$T : C^n(\mathbb{R}, X) \to C(\mathbb{R}, X)$$

defined by

$$T(u) = u^{(n)} + a_1 u^{(n-1)} + \ldots + a_n u, \qquad u \in C^n(\mathbb{R}, X),$$

where $a_1, \ldots, a_n \in \mathbb{C}$. We suppose that $C^n(\mathbb{R}, X)$ and $C(\mathbb{R}, X)$ are endowed with the same gauge $\|\cdot\|_\infty$.

Theorem 61 ([19]). *The operator T is Ulam stable if and only if its characteristic equation has not roots on the imaginary axis. Moreover, for every $\varepsilon > 0$, $u \in C^n(\mathbb{R}, X)$, and $v \in C(\mathbb{R}, X)$ such that*

$$\|Tu - v\|_\infty \leq \varepsilon,$$

there exists a unique $u_0 \in C^n(\mathbb{R}, X)$, $Tu_0 = v$ and

$$\|u - u_0\|_\infty \le \frac{\varepsilon}{\prod_{k=1}^{n} |\mathfrak{R}\lambda_k|}$$

where $\lambda_1, \ldots, \lambda_n \in \mathbb{C}$ are the roots of the characteristic equation of the operator T.

The previous results were concerned with differential operators defined on $C^n(\mathbb{R}, X)$. In the next section we consider differential operators defined on $C^n(I, X)$, where I is an arbitrary interval in $\mathbb{R}$, and obtain results on generalized Ulam stability. We also give sharp estimates of the involved constants.

2. Linear differential equation of the first order

In this section we present some results from [22].

In what follows, $I = (a, b)$, $a, b \in \mathbb{R} \cup \{\pm\infty\}$ is an open interval in $\mathbb{R}$, $c \in [a, b]$, $f \in C(I, X)$, $\lambda \in C(I, \mathbb{K})$, and $\varepsilon \in C(I, \mathbb{R})$ with $\varepsilon \ge 0$. We deal with the stability of the linear differential equation (see [22])

$$y'(x) - \lambda(x)y(x) = f(x), \qquad x \in I. \tag{3.1}$$

For a function $g : (a, b) \to X$ define

$$g(a) := \lim_{x \to a^+} g(x)$$

and

$$g(b) := \lim_{x \to b^-} g(x),$$

if the limits exist. Let $L \in C^1(I, \mathbb{K})$ be an antiderivative of λ, i.e., $L' = \lambda$ on I. Define $\psi_c : I \to \mathbb{R}$ by

$$\psi_c(x) := e^{\mathfrak{R}L(x)} \left| \int_c^x e^{-\mathfrak{R}L(t)} \varepsilon(t)dt \right|. \tag{3.2}$$

If $c = a$ or $c = b$, then we suppose that the integral, which defines ψ_c, is convergent for every $x \in I$. Therefore $\psi_c(c) = 0$ for all $c \in I$.

The following well known-lemma is useful in the proof of our stability results.

Lemma 1. *The general solution of the equation*

$$y'(x) - \lambda(x)y(x) = f(x), \qquad x \in I \tag{3.3}$$

is given by

$$y(x) = e^{L(x)} \left(\int_{x_0}^x f(t)e^{-L(t)}dt + k \right), \tag{3.4}$$

where $x_0 \in I$ and $k \in X$ is an arbitrary constant.

The first result on generalized stability for a first-order linear differential equation is contained in the next theorem.

Theorem 62. *For every $y \in C^1(I, X)$ satisfying*

$$\|y'(x) - \lambda(x)y(x) - f(x)\| \le \varepsilon(x), \qquad x \in I, \tag{3.5}$$

there exists a unique solution $u \in C^1(I, X)$ of equation (3.3) with the property

$$\|y(x) - u(x)\| \le \psi_c(x), \qquad x \in I. \tag{3.6}$$

Proof. Existence. Let $y \in C^1(I, X)$ satisfy (3.5) and define

$$g(x) := y'(x) - \lambda(x)y(x) - f(x), \qquad x \in I. \tag{3.7}$$

Then, according to Lemma 1, it follows

$$y(x) = e^{L(x)} \left(\int_{x_0}^x e^{-L(t)} f(t)dt + \int_{x_0}^x e^{-L(t)} g(t)dt + k \right), \qquad x_0 \in I, \ k \in X.$$

Let $G : I \to X$ be given by

$$G(x) := \int_c^x e^{-L(t)} g(t)dt, \qquad x \in I. \tag{3.8}$$

If $c = a$ or $c = b$, then the integral, which defines G, is convergent, because

$$\|g(t)\| \le \varepsilon(t) \text{ for all } t \in I.$$

(See the remark after (3.2)).

Now let u be defined by

$$u(x) := e^{L(x)} \left(\int_{x_0}^x f(t)e^{-L(t)}dt + k - G(x_0) \right).$$

Then obviously u satisfies equation (3.3) and we get

$$\|y(x) - u(x)\| = e^{\Re L(x)} \left\| \int_{x_0}^x g(t)e^{-L(t)}dt + G(x_0) \right\| = e^{\Re L(x)} \|G(x)\|$$

$$\le e^{\Re L(x)} \left| \int_c^x \|e^{-L(t)} g(t)\| dt \right|$$

$$\le e^{\Re L(x)} \left| \int_c^x e^{-\Re L(t)} \varepsilon(t)dt \right|$$

$$= \psi_c(x), \ x \in I.$$

Therefore the existence is proved.

Uniqueness. Suppose that for y satisfying (3.5) there exist u_1, u_2, $u_1 \neq u_2$, such that (3.3) and (3.6) are valid. Then

$$u_j(x) = e^{L(x)} \left(\int_{x_0}^{x} f(t)e^{-L(t)}dt + k_j \right), \qquad k_j \in X, \; j = 1, 2, \; k_1 \neq k_2$$

and

$$e^{\Re L(x)} \|k_1 - k_2\| = \|u_1(x) - u_2(x)\|$$
$$\leq \|u_1(x) - y(x)\| + \|y(x) - u_2(x)\|$$
$$\leq 2e^{\Re L(x)} \left| \int_{c}^{x} e^{-\Re L(t)} \varepsilon(t)dt \right|$$

for all $x \in I$. Therefore

$$\|k_1 - k_2\| \leq 2 \left| \int_{c}^{x} e^{-\Re L(t)} \varepsilon(t)dt \right|, \qquad x \in I. \tag{3.9}$$

Now, letting $x \to c$ in (3.9), we obtain $k_1 = k_2$, which is a contradiction. □

Theorem 62 leads to the following result concerning the Cauchy problem for equation (3.3).

Corollary 3. *Let $C \in X$. For every $y \in C^1(I, X)$ satisfying*

$$\begin{cases} \|y'(x) - \lambda(x)y(x) - f(x)\| \leq \varepsilon(x), \\ y(c) = C, \end{cases} \qquad x \in I$$

there exists a unique solution $u \in C^1(I, X)$ of the Cauchy problem

$$\begin{cases} u'(x) - \lambda(x)u(x) - f(x) = 0, \\ u(c) = C, \end{cases} \qquad x \in I,$$

with the property

$$\|y(x) - u(x)\| \leq \psi_c(x), \qquad x \in I.$$

The result obtained in Theorem 62 is more general than the result of [6, Lemma 2.2] and [12, Theorem 1], because it gives a better estimation of the difference between the approximate solution and the exact solution of equation (3.3). This is obvious in the cases $c = a$ and $c = b$, but for $c \in (a, b)$ this better approximation is not always valid on the entire interval (a, b). We will show in the next example that in some cases this estimation is global for $c \in (a, b)$, and we will find the optimal ψ_c.

Example 9. Let $\theta \in \mathbb{R} \setminus \{0\}$ and $\varepsilon(x) = \theta \mathfrak{R}\lambda(x)$ for $x \in I$. Then

$$\psi_c(x) = |\theta| \cdot |1 - e^{\mathfrak{R}(L(x)-L(c))}|, \qquad x \in I.$$

First we consider the case $\theta > 0$, i.e., $\mathfrak{R}\lambda(x) \geq 0$ for all $x \in I$. Then

$$\mathfrak{R}L'(x) \geq 0, \qquad x \in I,$$

hence $\mathfrak{R}L$ is increasing on I,

$$\psi_c(x) = \theta \cdot \begin{cases} e^{\mathfrak{R}L(x)-\mathfrak{R}L(c)} - 1, & x \in [c, b), \\ 1 - e^{\mathfrak{R}L(x)-\mathfrak{R}L(c)}, & x \in (a, c), \end{cases}$$

and

$$\|\psi_c\|_\infty = \theta \max\left\{e^{\mathfrak{R}(L(b)-L(c))} - 1, 1 - e^{\mathfrak{R}(L(a)-L(c))}\right\}.$$

Obviously $\|\psi_c\|_\infty$ is minimum for

$$e^{\mathfrak{R}(L(b)-L(c))} - 1 = 1 - e^{\mathfrak{R}(L(a)-L(c))},$$

i.e.,

$$e^{\mathfrak{R}L(c)} = \frac{e^{\mathfrak{R}L(a)} + e^{\mathfrak{R}L(b)}}{2}.$$

That equality determines the optimal constant $\widetilde{c}$; therefore the following estimation holds

$$\|y - u\|_\infty \leq \|\psi_{\widetilde{c}}\|_\infty,$$

where

$$\|\psi_{\widetilde{c}}\|_\infty = \theta(e^{\mathfrak{R}(L(b)-L(\widetilde{c}))} - 1) = \theta \cdot \frac{e^{\mathfrak{R}L(b)} - e^{\mathfrak{R}L(a)}}{e^{\mathfrak{R}L(b)} + e^{\mathfrak{R}L(a)}},$$

i.e.,

$$\min_{c \in I} \|\psi_c\|_\infty = \theta \cdot \frac{e^{\mathfrak{R}L(b)} - e^{\mathfrak{R}L(a)}}{e^{\mathfrak{R}L(b)} + e^{\mathfrak{R}L(a)}}. \tag{3.10}$$

The case $\theta < 0$ leads analogously to

$$\min_{c \in I} \|\psi_c\|_\infty = -\theta \cdot \frac{e^{\mathfrak{R}L(a)} - e^{\mathfrak{R}L(b)}}{e^{\mathfrak{R}L(b)} + e^{\mathfrak{R}L(a)}};$$

therefore for all $\theta \in \mathbb{R} \setminus \{0\}$ we have

$$\min_{c \in I} \|\psi_c\|_\infty = |\theta| \cdot \frac{|e^{\mathfrak{R}L(b)} - e^{\mathfrak{R}L(a)}|}{e^{\mathfrak{R}L(b)} + e^{\mathfrak{R}L(a)}}.$$

Remark 10. Now let λ be a constant with $\Re\lambda \neq 0$. Then $L(x) = \lambda x$ and

$$\min_{c \in I} \|\psi_c\|_\infty = |\theta| \cdot \frac{|e^{b\Re\lambda} - e^{a\Re\lambda}|}{e^{b\Re\lambda} + e^{a\Re\lambda}}. \tag{3.11}$$

Taking now an arbitrary $\delta > 0$ and $\theta = \dfrac{\delta}{|\Re\lambda|}$, it is easy to check that

$$\min_{c \in I} \|\psi_c\|_\infty < \frac{\delta}{|\Re\lambda|}(1 - e^{-|\Re\lambda|(b-a)})$$

if $a, b \in \mathbb{R}$, and

$$\min_{c \in I} \|\psi_c\|_\infty = \frac{\delta}{|\Re\lambda|}$$

if $a = -\infty$ or $b = +\infty$. Therefore we improve the result obtained in [6, Corollary 2.4] in the case of classical Ulam stability.

More precisely we have the following result:

Corollary 4. *Suppose that $\lambda \in \mathbb{C}$ with $\Re\lambda \neq 0$ and $\delta \geq 0$. Then for every $y \in C^1(I, X)$ satisfying*

$$\|y'(x) - \lambda y(x) - f(x)\| \leq \delta, \qquad x \in I$$

there exists a unique solution of (3.3) such that

$$\|y(x) - u(x)\| \leq \begin{cases} \dfrac{\delta}{|\Re\lambda|} \cdot \dfrac{|e^{b\Re\lambda} - e^{a\Re\lambda}|}{e^{b\Re\lambda} + e^{a\Re\lambda}}, & a, b \in \mathbb{R} \\[4mm] \dfrac{\delta}{|\Re\lambda|}, & a = -\infty \text{ or } b = +\infty. \end{cases}$$

3. Linear differential equation of a higher order with constant coefficients

The results presented in the previous section leads to the stability of the linear differential equation with constant coefficients (see [22]). In what follows we present some improvements of the results obtained in [19] and [6] for this equation. Suppose that $I = (a, b)$, with some $a, b \in \mathbb{R} \cup \{-\infty, +\infty\}$, $(X, \|\cdot\|)$ is a Banach space over $\mathbb{C}$, $f \in C(I, X)$, $\varepsilon \in C(I, \mathbb{R})$ and $\varepsilon \geq 0$, n is a positive integer, and $a_0, a_1, \ldots, a_{n-1} \in \mathbb{C}$ are given numbers. We study the stability of the linear differential equation:

$$y^{(n)}(x) - \sum_{j=0}^{n-1} a_j y^{(j)}(x) = f(x), \qquad x \in I. \tag{3.12}$$

Let

$$P(z) = z^n - \sum_{j=0}^{n-1} a_j z^j \tag{3.13}$$

be the characteristic polynomial of equation (3.12) and denote by $r_1, r_2, \ldots, r_n$ the complex roots of (3.13). For $\lambda \in \mathbb{C}$ and $c \in [a, b]$ define

$$\phi_\lambda(h)(x) := e^{\Re \lambda x} \left| \int_c^x e^{-\Re \lambda t} h(t) dt \right|, \qquad x \in I, \tag{3.14}$$

for all h with the property that the integral from the right-hand side of (3.14) is convergent. We suppose that $\phi_{r_k} \circ \phi_{r_{k-1}} \circ \ldots \circ \phi_{r_1}(\varepsilon)$ exist for every $k \in \{1, 2, \ldots, n\}$ if $c = a$ or $c = b$.

Theorem 63. *For every $y \in C^n(I, X)$ with the property*

$$\left\| y^{(n)}(x) - \sum_{j=0}^{n-1} a_j y^{(j)}(x) - f(x) \right\| \leq \varepsilon(x), \qquad x \in I \tag{3.15}$$

there exists a solution $u \in C^n(I, X)$ of equation (3.12) such that

$$\| y(x) - u(x) \| \leq \phi_{r_n} \circ \phi_{r_{n-1}} \circ \ldots \circ \phi_{r_1}(\varepsilon)(x), \qquad x \in I. \tag{3.16}$$

Proof. The proof by induction is analogous to the proof of [6, Theorem 2.3].

For $n = 1$, Theorem 63 holds by virtue of Theorem 62.

Now suppose that Theorem 63 holds for an $n \in \mathbb{N}$ (positive integers). We have to prove that for all $y \in C^{n+1}(I, X)$ satisfying the relation

$$\left\| y^{(n+1)}(x) - \sum_{j=0}^{n} a_j y^{(j)}(x) - f(x) \right\| \leq \varepsilon(x), \qquad x \in I, \tag{3.17}$$

there exists a solution $u \in C^{n+1}(I, X)$ satisfying

$$u^{(n+1)}(x) - \sum_{j=0}^{n} a_j u^{(j)}(x) - f(x) = 0, \qquad x \in I, \tag{3.18}$$

such that

$$\| y(x) - u(x) \| \leq \phi_{r_{n+1}} \circ \phi_{r_n} \circ \ldots \circ \phi_{r_1}(\varepsilon)(x), \qquad x \in I. \tag{3.19}$$

Let $y \in C^{n+1}(I, X)$ be a mapping satisfying (3.17). According to Vieta's relations we get

$$\| y^{(n+1)}(x) - (r_1 + \ldots + r_{n+1}) y^{(n)}(x) + \ldots + (-1)^{n+1} r_1 r_2 \ldots r_{n+1} y(x) - f(x) \| \leq \varepsilon(x)$$

or

$$\|(y^{(n+1)}(x) - r_{n+1}y^{(n)}(x)) - (r_1 + \ldots + r_n)(y^{(n)}(x) - r_{n+1}y^{(n-1)}(x)) + \ldots +$$

$$+(-1)^n r_1 \ldots r_n(y'(x) - r_{n+1}y(x)) - f(x)\| \le \varepsilon(x), \qquad x \in I. \qquad (3.20)$$

Let z be given by

$$z := y' - r_{n+1}y.$$

Then (3.20) becomes

$$\|z^{(n)}(x) - (r_1 + \ldots + r_n)z^{(n-1)}(x) + \ldots + (-1)^n r_1 \ldots r_n z(x) - f(x)\| \le \varepsilon(x)$$

for all $x \in I$. Therefore, by virtue of the induction hypothesis, there exists a mapping $v \in C^n(I, X)$ such that

$$v^{(n)}(x) - (r_1 + \ldots + r_n)v^{(n-1)}(x) + \ldots + (-1)^n r_1 \ldots r_n v(x) = f(x), \qquad x \in I,$$

and

$$\|z(x) - v(x)\| \le \phi_{r_n} \circ \ldots \circ \phi_{r_1}(\varepsilon)(x), \qquad x \in I,$$

which is equivalent to

$$\|y'(x) - r_{n+1}y(x) - v(x)\| \le \phi_{r_n} \circ \ldots \circ \phi_{r_1}(\varepsilon)(x).$$

Taking account of Theorem 62 it follows that there exists a unique mapping $u \in C^1(I, X)$ such that

$$u'(x) - r_{n+1}u(x) - v(x) = 0, \qquad x \in I, \qquad (3.21)$$

and

$$\|y(x) - u(x)\| \le \phi_{r_{n+1}} \circ \phi_{r_n} \circ \ldots \circ \phi_{r_1}(\varepsilon)(x), \qquad x \in I.$$

Finally, taking into account the properties of u and v, it follows that u belongs to $C^{n+1}(I, X)$ and satisfies (3.18). The theorem is proved. $\qquad \square$

Theorem 64. *Let δ be a positive number and suppose that all the roots of the characteristic equation (3.13) have the property $\Re r_k \neq 0$, $1 \le k \le n$. Then for every mapping $y \in C^n(I, X)$ satisfying the relation*

$$\left\| y^{(n)}(x) - \sum_{j=0}^{n-1} a_j y^{(j)}(x) - f(x) \right\| \le \delta, \qquad x \in I,$$

there exists a solution $u \in C^n(I, X)$ of the equation

$$u^{(n)}(x) - \sum_{j=0}^{n-1} a_j u^{(j)}(x) - f(x) = 0, \qquad x \in I,$$

such that

$$\|y(x) - u(x)\| \leq L,$$

where

$$L = \begin{cases} \delta \cdot \prod_{k=1}^{n} \dfrac{1}{|\Re r_k|} \cdot \dfrac{|e^{b\Re r_k} - e^{a\Re r_k}|}{e^{b\Re r_k} + e^{a\Re r_k}}, & \text{if} \quad a, b \in \mathbb{R}, \\[3ex] \dfrac{\delta}{\prod_{k=1}^{n} |\Re r_k|}, & \text{if} \quad a = -\infty \text{ or } b = +\infty. \end{cases}$$

Proof. The proof follows analogously to the proof of Theorem 63 taking account of Corollary 4. □

Remark 11. The uniqueness of the solution u in Theorem 64 holds if the characteristic polynomial P does not have pure imaginary roots and $I = \mathbb{R}$ (see [20]).

4. First-order linear differential operator

In what follows, $I = (a, b)$, $a, b \in \mathbb{R} \cup \{\pm\infty\}$ is an open interval, $c \in (a, b)$, $(X, \|\cdot\|)$ is a Banach space over $\mathbb{C}$, $C^n(I, X)$ is the set of all n-times strongly differentiable functions $f : I \to X$ with $f^{(n)}$ continuous on I, $n \in \mathbb{N}$, and $C(I, X)$ is the set of all continuous functions $f : I \to X$.

Also, let $\lambda, a_1, \ldots, a_n \in C(I, \mathbb{C})$ be given. We deal with the Ulam stability of the linear differential operator $D^n : C^n(I, X) \to C(I, X)$ defined by

$$D^n(y) = y^{(n)} + a_1 y^{(n-1)} + \ldots + a_n y, \qquad y \in C^n(I, X). \tag{3.22}$$

For every $h \in C^n(I, X)$ define $\|h\|_\infty$ by

$$\|h\|_\infty = \sup\{\|h(t)\| : t \in I\}. \tag{3.23}$$

Then $\|\cdot\|_\infty$ is a gauge function on $C^n(I, X)$ (for the definition of a gauge see Chapter 2). For an arbitrary function $f : A \to B$ we denote by $R(f)$ the range of f, i.e.,

$$R(f) = \{y \mid y = f(x), \ x \in A\}.$$

Definition 7. The operator D^n is said to be stable in the Ulam sense if for every $\varepsilon \geq 0$

there exists $\delta \geq 0$ such that for every $f \in R(D^n)$ and every $y \in C^n(I, X)$ satisfying

$$\|D^n(y) - f\|_\infty \leq \varepsilon \tag{3.24}$$

there exists $u \in C^n(I, X)$ such that $D^n(u) = f$ and

$$\|y - u\|_\infty \leq \delta. \tag{3.25}$$

Let ε be a nonnegative number. As in Section 2, for a function $g : (a, b) \to X$ define

$$g(a) := \lim_{x \to a+0} g(x), \qquad g(b) := \lim_{x \to b-0} g(x),$$

if the limits exist. Let L be an antiderivative of λ, i.e., $L \in C^1(I, \mathbb{C})$ and $L' = \lambda$ on I. For $n = 1$ and $a_1 = \lambda$ denote D^1 by D_λ, i.e.

$$D_\lambda = y' + \lambda y, \qquad y \in C^1(I, X).$$

The following results concern the Ulam stability of the first-order linear differential operator (see [23]).

Theorem 65. *Suppose that*

$$\inf_{x \in I} |\mathfrak{R}\lambda(x)| := m > 0. \tag{3.26}$$

Then for every $f \in C(I, X)$ and every $y \in C^1(I, X)$ satisfying

$$\|D_\lambda(y) - f\|_\infty \leq \varepsilon \tag{3.27}$$

there exists $u \in C^1(I, X)$ with the properties $D_\lambda(u) = f$ and

$$\|y - u\|_\infty \leq \frac{\varepsilon}{m} \cdot \delta_\lambda \tag{3.28}$$

where

$$\delta_\lambda = \begin{cases} 1 - e^{\mathfrak{R}L(a) - \mathfrak{R}L(b)}, & \text{if } \mathfrak{R}\lambda > 0 \quad \text{on} \quad I, \\ 1 - e^{\mathfrak{R}L(b) - \mathfrak{R}L(a)}, & \text{if } \mathfrak{R}\lambda < 0 \quad \text{on} \quad I. \end{cases}$$

Moreover, if one of the following conditions,

$$\begin{aligned} &i) \quad \mathfrak{R}L(a) = -\infty, \quad \text{if } \mathfrak{R}\lambda > 0 \quad \text{on} \quad I, \\ &ii) \quad \mathfrak{R}L(b) = -\infty, \quad \text{if } \mathfrak{R}\lambda < 0 \quad \text{on} \quad I, \end{aligned} \tag{3.29}$$

is satisfied, then u is uniquely determined.

Proof. From (3.26) it follows that $\mathfrak{R}\lambda \neq 0$ on I; therefore, $\mathfrak{R}\lambda$ has a constant sign on I, in view of its continuity. We conclude that $\mathfrak{R}L$ is strictly monotone on I, hence there exist $\mathfrak{R}L(a)$ and $\mathfrak{R}L(b)$, finite or infinite.

Existence. Let $y \in C^1(I, X)$ satisfying (3.27) and define

$$g(x) := y'(x) + \lambda(x)y(x) - f(x), \qquad x \in I.$$

Then, according to Lemma 1, we get

$$y(x) = e^{-L(x)}\left(\int_{x_0}^x e^{L(t)} f(t)dt + \int_{x_0}^x e^{L(t)} g(t)dt + k\right), \qquad x_0 \in I, \ k \in X. \qquad (3.30)$$

$1°$ Suppose first that $\mathfrak{R}\lambda > 0$ on I. Define

$$G(x) := \int_a^x e^{L(t)} g(t)dt, \qquad x \in I.$$

Since I is an open interval we have to prove that $G(x)$ is defined for all $x \in I$. We get

$$\|e^{L(t)} g(t)\| \le \varepsilon \cdot e^{\mathfrak{R}L(t)}, \qquad t \in I, \qquad (3.31)$$

and

$$\int_a^x e^{\mathfrak{R}L(t)}dt = \int_a^x \frac{1}{\mathfrak{R}\lambda(t)} \cdot \mathfrak{R}\lambda(t) \cdot e^{\mathfrak{R}L(t)}dt \le \frac{1}{m}\int_a^x (e^{\mathfrak{R}L(t)})'dt$$
$$= \frac{1}{m}(e^{\mathfrak{R}L(x)} - e^{\mathfrak{R}L(a)}) \le \frac{1}{m}e^{\mathfrak{R}L(x)}, \qquad x \in I. \qquad (3.32)$$

($\mathfrak{R}L(a) < \mathfrak{R}L(x)$ for all $x \in I$, since $\mathfrak{R}L$ is increasing).

From (3.31) and (3.32) it follows that $G(x)$ is absolutely convergent for all $x \in I$. Now defining

$$u(x) := e^{-L(x)}\left(\int_{x_0}^x e^{L(t)} f(t)dt + k - G(x_0)\right), \qquad x \in I,$$

and using (3.32) we get

$$\|y(x) - u(x)\| \le \varepsilon e^{-\mathfrak{R}L(x)} \int_a^x e^{\mathfrak{R}L(t)}dt$$
$$\le \frac{\varepsilon}{m}e^{-\mathfrak{R}L(x)}(e^{\mathfrak{R}L(x)} - e^{\mathfrak{R}L(a)})$$
$$\le \frac{\varepsilon}{m}(1 - e^{\mathfrak{R}L(a)-\mathfrak{R}L(b)}), \qquad x \in I.$$

$2°$ The case $\mathfrak{R}\lambda < 0$ can be treated analogously, setting

$$G(x) := -\int_x^b e^{L(t)} g(t)dt, \qquad x \in I.$$

The existence is proved.

Uniqueness. Suppose that one of the conditions, i) or ii), is satisfied and for a function $y \in C^1(I, X)$ satisfying (3.27) there exist two solutions u_1, u_2 of $D_\lambda(u) = f$, $u_1 \ne u_2$, with the property (3.28). Then

$$u_j(x) = e^{-L(x)}\left(\int_{x_0}^x e^{L(t)}f(t)dt + k_j\right), \qquad k_j \in X, \; j = 1, 2,$$

with $k_1 \neq k_2$, according to Lemma 1, and

$$e^{-\Re L(x)}\|k_1 - k_2\| = \|u_1(x) - u_2(x)\|$$
$$\leq \|u_1(x) - y(x)\| + \|y(x) - u_2(x)\|$$
$$\leq \frac{2\varepsilon}{m}\delta_\lambda, \qquad x \in I. \tag{3.33}$$

Letting in (3.33) $x \to a$ if i) is satisfied, or $x \to b$ if ii) is satisfied, it follows that $\infty \leq \dfrac{2\varepsilon}{m}$, which is a contradiction.

The theorem is proved. □

Theorem 66. *Suppose that* $\inf_{x \in I}|\Re\lambda(x)| := m > 0$ *and let* $f \in C(I, X)$. *Then for every* $y \in C^1(I, X)$ *satisfying*

$$\|D_\lambda(y) - f\|_\infty \leq \varepsilon \tag{3.34}$$

there exists $u \in C^1(I, X)$ *with the properties* $D_\lambda(u) = f$ *and*

$$\|y - u\|_\infty \leq \frac{\varepsilon}{m}\delta_\lambda(c), \tag{3.35}$$

where

$$\delta_\lambda(c) = \begin{cases} \mu_1, & \text{if } \Re\lambda > 0 \text{ and } \Re L(a) > -\infty, \\ \mu_2, & \text{if } \Re\lambda < 0 \text{ and } \Re L(b) > -\infty, \\ 1, & \text{if } \Re L(a) = -\infty \text{ or } \Re L(b) = -\infty, \end{cases} \tag{3.36}$$

$$\mu_1 := \max\{1 - e^{\Re L(c) - \Re L(b)}, e^{\Re L(c) - \Re L(a)} - 1\},$$

$$\mu_2 := \max\{e^{\Re L(c) - \Re L(b)} - 1, 1 - e^{\Re L(c) - \Re L(a)}\}.$$

Proof. If $\Re L(a) = -\infty$ or $\Re L(b) = -\infty$, the statement follows from Theorem 65.

Suppose now that $\Re L(a) > -\infty$ and $\Re L(b) > -\infty$. Similarly to the proof of Theorem 65, we get that $\Re\lambda$ has a constant sign on I and $\Re L$ is strictly monotone on I.

Let $y \in C^1(I, X)$ satisfying (3.34) and

$$g(x) := y'(x) + \lambda(x)y(x) - f(x), \qquad x \in I.$$

Then y is given by (3.30). Define G and u by

$$G(x) := \int_c^x e^{L(t)} g(t) dt$$

$$u(x) := e^{-L(x)} \left(\int_{x_0}^x e^{L(t)} f(t) dt + k - G(x_0) \right), \qquad x \in I.$$

We get, analogously to the proof of Theorem 65,

$$\|y(x) - u(x)\| = e^{-\Re L(x)} \|G(x)\|$$
$$\leq e^{-\Re L(x)} \left| \int_c^x \|e^{L(t)} g(t)\| dt \right|$$
$$\leq \varepsilon e^{-\Re L(x)} \left| \int_c^x e^{\Re L(t)} dt \right|, \qquad x \in I. \tag{3.37}$$

On the other hand, since $\Re \lambda \cdot e^{\Re L}$ has constant sign on I, it follows

$$\left| \int_c^x e^{\Re L(t)} dt \right| = \left| \int_c^x \frac{1}{\Re \lambda(t)} \cdot \Re \lambda(t) \cdot e^{\Re L(t)} dt \right|$$
$$\leq \frac{1}{m} |e^{\Re L(x)} - e^{\Re L(c)}|, \qquad x \in I. \tag{3.38}$$

The relations (3.37) and (3.38) lead to

$$\|y(x) - u(x)\| \leq \frac{\varepsilon}{m} |1 - e^{\Re L(c) - \Re L(x)}|, \qquad x \in I. \tag{3.39}$$

Moreover, (3.35) follows from (3.39) taking account of the monotonicity of $\Re L$.
The theorem is proved. $\qquad\qquad\qquad\qquad\qquad\qquad\qquad\qquad\qquad\qquad\qquad\qquad\square$

Remark 12. If $L(a) > -\infty$ and $L(b) > -\infty$ it is easy to verify that $\delta_\lambda(c)$ is minimal in Theorem 66 for

$$\delta_\lambda(c) = \begin{cases} 1 - e^{\Re L(c) - \Re L(b)} = e^{\Re L(c) - \Re L(a)} - 1, & \text{if } \Re \lambda > 0, \\ e^{\Re L(c) - \Re L(b)} - 1 = 1 - e^{\Re L(c) - \Re L(a)}, & \text{if } \Re \lambda < 0. \end{cases} \tag{3.40}$$

The above two cases give

$$e^{-\Re L(c)} = \frac{e^{-\Re L(a)} + e^{-\Re L(b)}}{2} \tag{3.41}$$

The relation (3.41) determines the optimal constant $\widetilde{c}$. Note that since $\Re L$ is strictly monotone and continuous on I, $\widetilde{c}$ exists and is unique.
Choosing $c = \widetilde{c}$ in Theorem 66 we get

$$\delta_\lambda(\widetilde{c}) = \frac{|e^{-\Re L(a)} - e^{-\Re L(b)}|}{e^{-\Re L(a)} + e^{-\Re L(b)}}. \tag{3.42}$$

Therefore, the relation (3.35) becomes

$$\|y - u\|_\infty \leq \frac{\varepsilon}{m}\delta_\lambda(\widetilde{c}). \tag{3.43}$$

5. Higher-order linear differential operator

The results expressed in Theorem 65 and Theorem 66 lead to the Ulam stability of the operator D^n defined by (3.22), in appropriate conditions. We suppose that there exist $r_1, r_2, \ldots, r_n \in C(I, \mathbb{C})$ such that

$$D^n = D_{r_1} \circ D_{r_2} \circ \ldots \circ D_{r_n}. \tag{3.44}$$

We remark that D^n is a surjective operator as a composition of surjective operators (D_λ is a surjective operator in view of Lemma 1).

Let R_k be an antiderivative of r_k, with $1 \leq k \leq n$, and $f \in C(I, X)$ be an arbitrary function.

Theorem 67. *Suppose that*

$$\inf_{x \in I} |r_k(x)| := m_k > 0, \qquad k \in \{1, 2, \ldots, n\}.$$

Then for every $y \in C^n(I, X)$ satisfying the relation

$$\|D^n(y) - f\|_\infty \leq \varepsilon \tag{3.45}$$

there exists $u \in C^n(I, X)$ with the properties $D^n(u) = f$ and

$$\|y - u\|_\infty \leq \frac{\varepsilon}{m_1 m_2 \ldots m_n}\delta_{r_1}\delta_{r_2}\ldots\delta_{r_n}. \tag{3.46}$$

Proof. We prove the theorem by induction on n.

For $n = 1$, Theorem 67 holds by virtue of Theorem 65.

Now suppose that Theorem 67 holds for an $n \in \mathbb{N}$. We have to prove that for all $y \in C^{n+1}(I, X)$ satisfying

$$\|D^{n+1}(y) - f\|_\infty \leq \varepsilon \tag{3.47}$$

there exists $u \in C^{n+1}(I, X)$, $D^{n+1}(u) = f$, such that

$$\|y - u\|_\infty \leq \frac{\varepsilon}{m_1 m_2 \ldots m_{n+1}}\delta_{r_1}\delta_{r_2}\ldots\delta_{r_{n+1}}. \tag{3.48}$$

Let $y \in C^{n+1}(I, X)$ satisfying (3.47). Then

$$\|D^n(z) - f\|_\infty \leq \varepsilon$$

with $z := D_{r_{n+1}}(y)$. Hence, by virtue of the induction hypothesis, there exists $v \in$

$C^n(I, X)$, $D^n(v) = f$, and

$$\|z - v\|_\infty \leq \frac{\varepsilon}{m_1 m_2 \ldots m_n} \delta_{r_1} \delta_{r_2} \ldots \delta_{r_n},$$

which is equivalent to

$$\|D_{r_{n+1}}(y) - v\|_\infty \leq \frac{\varepsilon}{m_1 m_2 \ldots m_n} \delta_{r_1} \delta_{r_2} \ldots \delta_{r_n}. \tag{3.49}$$

Then, according to Theorem 65, from (3.49) it follows that there exists a mapping $u \in C^1(I, X)$, $D_{r_{n+1}}(u) = v$, and

$$\|y - u\|_\infty \leq \frac{\varepsilon}{m_1 m_2 \ldots m_{n+1}} \delta_{r_1} \delta_{r_2} \ldots \delta_{r_{n+1}}.$$

Finally, the relations

$$v \in C^n(I, X), \qquad D^n(v) = f, \qquad D_{r_{n+1}}(u) = v$$

lead to

$$u \in C^{n+1}(I, X), \qquad D^{n+1}(u) = f.$$

The theorem is proved. □

An analogous result follows from Theorem 66.

Theorem 68. *Suppose that* $\inf\limits_{x \in I} |r_k(x)| = m_k > 0$ *for every* $k \in \{1, 2, \ldots, n\}$. *Then for every* $y \in C^n(I, X)$ *satisfying*

$$\|D^n(y) - f\|_\infty \leq \varepsilon$$

there exists $u \in C^n(I, X)$, $D^n(u) = f$, *such that*

$$\|y - u\|_\infty \leq \frac{\varepsilon}{m_1 m_2 \ldots m_n} \delta_{r_1}(c) \ldots \delta_{r_n}(c). \tag{3.50}$$

Proof. Analogous to the proof of Theorem 67. □

Remark 13. If $\mathfrak{R} R_k(a) > -\infty$ and $\mathfrak{R} R_k(b) > -\infty$ for all $k \in \{1, 2, \ldots, n\}$, choosing $c = \widetilde{c}$ in Theorem 68, we obtain that estimate (3.50) becomes

$$\|y - u\|_\infty \leq \frac{\varepsilon}{m_1 m_2 \ldots m_n} \prod_{k=1}^{n} \frac{|e^{-\mathfrak{R} R_k(a)} - e^{-\mathfrak{R} R_k(b)}|}{e^{-\mathfrak{R} R_k(a)} + e^{-\mathfrak{R} R_k(b)}}.$$

Proof. Follows from Theorem 68 and Remark 12. □

The results obtained in Theorem 67, Theorem 68, and their consequences improve and extend the estimates given in [6, Corollary 2.5] and [22, Theorem 3.2].

6. Partial differential equations

In what follows let $D = [a, b) \times \mathbb{R}$, $a \in \mathbb{R}$, $b \in \mathbb{R} \cup \{+\infty\}$ be a subset of $\mathbb{R}^2$. We deal with the Ulam stability of the linear partial differential equation

$$p(x, y)\frac{\partial u}{\partial x} + q(x, y)\frac{\partial u}{\partial y} = p(x, y)r(x)u + f(x, y), \tag{3.51}$$

where $p, q \in C(D, \mathbb{K})$, $f \in C(D, X)$, $r \in C([a, b), \mathbb{R})$ are given functions and $u \in C^1(D, X)$ is the unknown function (see [16]). We suppose that $p(x, y) \neq 0$ for every $(x, y) \in D$.

Let $\varepsilon \geq 0$ be a given number. The equation (3.51) is said to be stable in Ulam sense if there exists $\delta \geq 0$ such that for every function $u \in C^1(D, X)$ satisfying

$$\left\| p(x, y)\frac{\partial u}{\partial x}(x, y) + q(x, y)\frac{\partial u}{\partial y}(x, y) - p(x, y)r(x)u(x, y) - f(x, y) \right\| \leq \varepsilon \tag{3.52}$$

for all $(x, y) \in D$ there exists a solution $v \in C^1(D, X)$ of (3.51) with the property

$$\|u(x, y) - v(x, y)\| \leq \delta, \qquad (x, y) \in D. \tag{3.53}$$

We will prove in what follows that the existence of a global prime integral $\varphi : [a, b) \to \mathbb{R}$ of equation (3.51) leads, in appropriate conditions, to the stability of equation (3.51). The following lemma is a useful tool in the proof of the main result of this section.

Lemma 2. *Let $\varphi : [a, b) \to \mathbb{R}$ be a solution of the differential equation*

$$y' = \frac{q(x, y)}{p(x, y)}.$$

Then u is a solution of equation (3.51) if and only if there exists a function $F \in C^1(I, X)$ such that

$$u(x, y) = e^{-L(x)}\left(\int_a^x \frac{f(\theta, \varphi(\theta) + y - \varphi(x))}{p(\theta, \varphi(\theta) + y - \varphi(x))} e^{L(\theta)} d\theta + F(y - \varphi(x)) \right) \tag{3.54}$$

for every $(x, y) \in D$, where $L(x) = -\int_a^x r(\theta)d\theta$ for $x \in [a, b)$ and

$$I = \{y - \varphi(x) : (x, y) \in D\}.$$

Proof. Let u be a solution of equation (3.51) and consider the change of coordinates

$$\begin{cases} s = x \\ t = y - \varphi(x) \end{cases} \Leftrightarrow \begin{cases} x = s \\ y = \varphi(s) + t. \end{cases} \tag{3.55}$$

Define the function v by

$$v(s, t) = u(s, \varphi(s) + t) \Leftrightarrow u(x, y) = v(x, y - \varphi(x)). \tag{3.56}$$

Then

$$\frac{\partial u}{\partial x} = \frac{\partial v}{\partial s} - \varphi'(s) \cdot \frac{\partial v}{\partial t}, \qquad \frac{\partial u}{\partial y} = \frac{\partial v}{\partial t}.$$

The substitution in (3.51) yields

$$\frac{\partial v}{\partial s} - r(s) \cdot v = \frac{f(s, \varphi(s) + t)}{p(s, \varphi(s) + t)}. \tag{3.57}$$

Equation (3.57) is equivalent to

$$\frac{\partial}{\partial s}(v \cdot e^{L(s)}) = \frac{f(s, \varphi(s) + t)}{p(s, \varphi(s) + t)} \cdot e^{L(s)}. \tag{3.58}$$

An integration on the interval $[a, s)$, $s \in [a, b)$, leads to

$$v(s, t) = e^{-L(s)} \left(\int_a^s \frac{f(\theta, \varphi(\theta) + t)}{p(\theta, \varphi(\theta) + t)} e^{L(\theta)} d\theta + F(t) \right), \tag{3.59}$$

where F is an arbitrary function of class C^1.

Replacing s, t from (3.55) in (3.59) the relation (3.54) is obtained.

Now let u be given by (3.54); we have to prove that u is a solution of (3.51). Taking account of the change of coordinates (3.55), it is sufficient to prove that v, given by (3.59), satisfies (3.57). A simple calculation shows that v is a solution of (3.57). □

The main result of this section is given in the next theorem.

Theorem 69. *Let $\varepsilon \geq 0$ be a given number. Suppose that the equation*

$$y' = \frac{q(x, y)}{p(x, y)}$$

admits a solution $\varphi : [a, b) \to \mathbb{R}$ and

$$\inf_{(x,y) \in D} |p(x, y)| \cdot r(x) =: m > 0.$$

Then for every solution u of (3.52) there exists a solution v of (3.51) with the property

$$\|u(x, y) - v(x, y)\| \leq \frac{\varepsilon}{m}, \qquad (x, y) \in D. \tag{3.60}$$

Moreover, if

$$L(b) =: \lim_{x \to b} L(x) = -\infty,$$

then v is uniquely determined.

Proof. Existence. Let u be a solution of (3.52) and put

$$p(x, y)\frac{\partial u}{\partial x}(x, y) + q(x, y)\frac{\partial u}{\partial y}(x, y) - p(x, y)r(x)u(x, y) - f(x, y) =: g(x, y)$$

for every $(x, y) \in D$. Then, according to Lemma 2, we have the following:

$$u(x, y) = e^{-L(x)}\left(\int_a^x \frac{f(\theta, \varphi(\theta) + y - \varphi(x)) + g(\theta, \varphi(\theta) + y - \varphi(x))}{p(\theta, \varphi(\theta) + y - \varphi(x))} e^{L(\theta)} d\theta \right.$$

$$\left. +F(y - \varphi(x)) \right),$$

where $F \in C^1(I, X)$ is an arbitrary function.
    Let v be defined by

$$v(x, y) = e^{-L(x)}\left(\int_a^x \frac{f(\theta, \varphi(\theta) + y - \varphi(x))}{p(\theta, \varphi(\theta) + y - \varphi(x))} e^{L(\theta)} d\theta \right.$$

$$\left. + \int_a^b \frac{g(\theta, \varphi(\theta) + y - \varphi(x))}{p(\theta, \varphi(\theta) + y - \varphi(x))} e^{L(\theta)} d\theta + F(y - \varphi(x)) \right), \qquad (x, y) \in D.$$

The function v is well defined since the integral

$$G(t) := \int_a^b \frac{g(\theta, \varphi(\theta) + t)}{p(\theta, \varphi(\theta) + t)} e^{L(\theta)} d\theta, \qquad t \in I$$

is convergent. Indeed,

$$\| G(t) \| \le \int_a^b \left\| \frac{g(\theta, \varphi(\theta) + t)}{p(\theta, \varphi(\theta) + t) \cdot r(\theta)} \cdot r(\theta)e^{L(\theta)} \right\| d\theta$$

$$\le \frac{\varepsilon}{m} \int_a^b r(\theta)e^{L(\theta)} d\theta$$

$$= -\frac{\varepsilon}{m} \int_a^b (e^{L(\theta)})' d\theta = \frac{\varepsilon}{m}(1 - e^{L(b)}) \le \frac{\varepsilon}{m}, \qquad t \in I,$$

therefore, $G(t)$ is absolutely convergent.
    (Since r is positive on $[a, b)$ it follows that the function L is decreasing on $[a, b)$; a

monotone function has left and right limits at every point, therefore

$$L(b) = -\lim_{x \to b} \int_a^x r(\theta)d\theta$$

exists and is negative).

On the other hand, v is a solution of (3.51) being of the form (3.54). We have the following:

$$\|u(x, y) - v(x, y)\| = \left\| e^{-L(x)}\left(-\int_x^b \frac{g(\theta, \varphi(\theta) + y - \varphi(x))}{p(\theta, \varphi(\theta) + y - \varphi(x))} \cdot e^{L(\theta)}d\theta \right) \right\|$$

$$\leq e^{-L(x)}\int_x^b \frac{\varepsilon}{|p(\theta, \varphi(\theta) + y - \varphi(x))|}e^{L(\theta)}d\theta$$

$$= e^{-L(x)}\int_x^b \frac{\varepsilon}{|p(\theta, \varphi(\theta) + y - \varphi(x))|r(\theta)}r(\theta)e^{L(\theta)}d\theta$$

$$\leq \frac{\varepsilon}{m}e^{-L(x)}\int_x^b (-e^{L(\theta)})'d\theta$$

$$= \frac{\varepsilon}{m}(1 - e^{L(b)-L(x)}) \leq \frac{\varepsilon}{m}, \qquad (x, y) \in D.$$

Uniqueness. Suppose that $L(b) = -\infty$ and for a solution u of (3.52) there exist two solutions v_1, v_2 of (3.51), $v_1 \neq v_2$, with the property (3.60), given by

$$v_k(x, y) = e^{-L(x)}\left(\int_a^x \frac{f(\theta, \varphi(\theta) + y - \varphi(x))}{p(\theta, \varphi(\theta) + y - \varphi(x))}e^{L(\theta)}d\theta + F_k(y - \varphi(x)) \right)$$

$(x, y) \in D, k \in \{1, 2\}$. We have

$$\|v_1(x, y) - v_2(x, y)\| \leq \|v_1(x, y) - u(x, y)\| + \|u(x, y) - v_2(x, y)\|$$

$$\leq \frac{2\varepsilon}{m}, \qquad (x, y) \in D,$$

which is equivalent to

$$e^{-L(x)}\|F_1(y - \varphi(x)) - F_2(y - \varphi(x))\| \leq \frac{2\varepsilon}{m}, \qquad (x, y) \in D. \qquad (3.61)$$

Since $v_1 \neq v_2$, it follows that there exists x_0 such that $F_1(x_0) \neq F_2(x_0)$. For $y = \varphi(x) + x_0$ the relation (3.61) becomes

$$e^{-L(x)}\|F_1(x_0) - F_2(x_0)\| \leq \frac{2\varepsilon}{m}, \qquad x \in [a, b). \qquad (3.62)$$

Now letting $x \to b$ in (3.62) it follows $\infty \leq \frac{2\varepsilon}{m}$, which is a contradiction. Uniqueness is proved. $\qquad\qquad \square$

Corollary 5. *Let* $D = (0, \infty) \times \mathbb{R}$ *and* $p, q \in C(D, \mathbb{R})$, $r \in C([0, \infty), \mathbb{R})$, $f \in C(D, X)$. *Suppose that* p, q *are homogeneous functions of the same degree,*

$$\frac{q(x, y)}{p(x, y)} \neq \frac{y}{x}$$

on D *and*

$$\inf_{(x,y) \in D} |p(x, y)| \cdot r(x) = m > 0.$$

Then for every $\varepsilon \geq 0$ *and every solution* u *of* (3.52) *there exists a solution* v *of* (3.51) *with the property* (3.53). *If*

$$\int_0^\infty r(\theta) d\theta = \infty,$$

then v *is uniquely determined.*

Proof. Suppose that p, q are homogeneous functions of the n-th degree. First we prove that the equation

$$y' = \frac{q(x, y)}{p(x, y)} \tag{3.63}$$

admits a solution $\varphi : (0, \infty) \to \mathbb{R}$.

Taking account of the homogeneity of p and q it follows

$$\frac{q(x, y)}{p(x, y)} = \frac{q\left(x \cdot 1, x \cdot \frac{y}{x}\right)}{p\left(x \cdot 1, x \cdot \frac{y}{x}\right)} = \frac{x^n q\left(1, \frac{y}{x}\right)}{x^n p\left(1, \frac{y}{x}\right)} = \frac{q\left(1, \frac{y}{x}\right)}{p\left(1, \frac{y}{x}\right)} =: h\left(\frac{y}{x}\right)$$

for all $(x, y) \in D$; therefore equation (3.63) is equivalent to the homogeneous differential equation

$$y' = h\left(\frac{y}{x}\right). \tag{3.64}$$

Let $H : \mathbb{R} \to \mathbb{R}$ be given by

$$H(z) = \int_0^z \frac{d\theta}{h(\theta) - \theta}, \qquad z \in \mathbb{R}. \tag{3.65}$$

Obviously H is well defined since $h(\theta) \neq \theta$ for all $\theta \in \mathbb{R}$.
The change of variable in (3.64) given by

$$y(x) = xz(x), \qquad x \in (0, \infty)$$

leads to the equation with separate variables

$$\frac{dz}{h(z) - z} = \frac{dx}{x}$$

with a solution given by

$$H(z) = \ln x, \qquad x \in (0, \infty). \tag{3.66}$$

By the condition $h(\theta) \neq \theta$, $\theta \in \mathbb{R}$, and the continuity of h, it follows that $h(\theta) - \theta$ has a constant sign on $\mathbb{R}$; therefore H is strictly monotone.

In this case there exists $H^{-1} : H(\mathbb{R}) \to \mathbb{R}$. From (3.66) we get the explicit solution of equation (3.64) given by

$$z(x) = H^{-1}(\ln x)$$

and finally the prime integral

$$\varphi(x) = x \cdot H^{-1}(\ln x), \qquad x \in (0, \infty).$$

Now the conclusion follows from Theorem 69. $\square$

Remark 14. If $m = 0$, then the result obtained in Theorem 69 is not generally true. Indeed, consider the equation

$$x\frac{\partial u}{\partial x} + y\frac{\partial u}{\partial y} = 0, \qquad x, y \in [a, \infty),\, a > 0, \tag{3.67}$$

and let $\varepsilon > 0$. A solution of the equation

$$x\frac{\partial u}{\partial x} + y\frac{\partial u}{\partial y} = \varepsilon$$

is of the form

$$u(x, y) = \varepsilon \ln x + \varphi\left(\frac{y}{x}\right),$$

where $\varphi : (0, \infty) \longrightarrow X$ is an arbitrary function of class C^1, according to Lemma 2.

Let

$$v(x, y) = \psi\left(\frac{y}{x}\right)$$

be an arbitrary solution of (3.67), $\psi \in C^1((0, \infty), X)$. The condition

$$\left\| x\frac{\partial u}{\partial x}(x, y) + y\frac{\partial u}{\partial y}(x, y) \right\| \le \varepsilon$$

is satisfied for all $x, y \in (0, \infty)$, but

$$\sup_{x \in [a, \infty)} \|u(x, x) - v(x, x)\| = +\infty,$$

and therefore equation (3.67) is not stable.

7. Laplace operator

In this section we investigate the Ulam stability of the Laplace operator Δ acting on a certain space of functions. The Ulam stability of Δ on other spaces was studied with different methods in [5], [8], and [10].

We investigate also the Ulam stability of the operator $p(x)\Delta$, where

$$p(x) := \frac{1}{2n}(1 - x_1^2 - \ldots - x_n^2)$$

for all x in the unit ball of $\mathbb{R}^n$. This operator is related to the infinitesimal generator of a C_0-semigroup systematically studied in [3], [4], [26] and the references given there. Moreover, we give explicit forms of the involved Ulam constants for Δ and $p(x)\Delta$.

Let $G \subset \mathbb{R}^n$, $n \geq 1$, be an open and connected set bounded by a surface S of class C^1.

Consider the Laplace operator,

$$\Delta = \frac{\partial^2}{\partial x_1^2} + \ldots + \frac{\partial^2}{\partial x_n^2},$$

acting on the space

$$D(\Delta) := \left\{ u \in C^2(G) \cap C(\overline{G}) : \Delta u \in C^1(G) \cap C(\overline{G}) \right\}.$$

On $C(\overline{G})$ and on its subspaces we consider the supremum norm denoted by $\|.\|$. In our approach we need the following result; for more details see [25, p. 68].

Theorem 70. *For each $f \in C^1(G) \cap C(\overline{G})$ there exists a unique $u \in D(\Delta)$ such that $\Delta u = f$ and $u|_S = 0$.*

Consequently $\Delta : D(\Delta) \to R(\Delta)$ is surjective, where the range of Δ is

$$R(\Delta) = C^1(G) \cap C(\overline{G}).$$

Moreover, according to Theorem 70 there exists a unique $q \in D(\Delta)$ such that

$$\begin{cases} \Delta q = -1, \\ q|_S = 0. \end{cases}$$

Theorem 71. *The operator $\Delta : D(\Delta) \to R(\Delta)$ is Ulam stable with constant $\|q\|$.*

Proof. Let $f \in D(\Delta)$, $\|\Delta f\| \le 1$. Then $\Delta f \in C^1(G) \cap C(\overline{G})$, and Theorem 70 guarantees the existence of $u \in D(\Delta)$ such that $\Delta u = \Delta f$ and $u|_S = 0$.

Then $\|\Delta u\| \le 1$, which entails $\Delta(u + q) = \Delta u - 1 \le 0$. Since $(u + q)|_S = 0$, we get $u + q \ge 0$. Similarly, $\Delta(u - q) = \Delta u + 1 \ge 0$ and $(u - q)|_S = 0$ imply $u - q \le 0$.

So we have $-q \le u \le q$, i.e., $\|u\| \le \|q\|$.

Let $g := f - u$. Then $\Delta g = 0$ and

$$\|f - g\| = \|u\| \le \|q\|,$$

which concludes the proof. $\square$

In particular, if G is the open unit ball of $\mathbb{R}^n$, then

$$q(x) = \frac{1}{2n}(1 - x_1^2 - \ldots - x_n^2)$$

and

$$\|q\| = \frac{1}{2n}.$$

Consequently we have the following:

Corollary 6. *If G is the open unit ball of $\mathbb{R}^n$, then $\Delta : D(\Delta) \to R(\Delta)$ is Ulam stable with constant $\frac{1}{2n}$.*

In what follows B will be the open unit ball of $\mathbb{R}^n$ with $n \ge 1$. Consider the function

$$p(x) = \frac{1}{2n}(1 - x_1^2 - \ldots - x_n^2), \qquad x \in \overline{B},$$

and the operator $W := p\Delta$ with domain

$$D(W) := \left\{ u \in C^2(B) \cap C(\overline{B}) : \Delta u \in C^1(B) \cap C(\overline{B}) \right\}.$$

According to Theorem 70, the range of W is

$$R(W) = \left\{ pv : v \in C^1(B) \cap C(\overline{B}) \right\}.$$

Define the function $h_n : [0, 1) \to \mathbb{R}$ by

$$h_n(t) := \frac{2n}{t^{n-1}} \int_0^t \frac{s^{n-1}}{1 - s^2} ds, \qquad t \in (0, 1),$$

and $h_n(0) := \lim_{t \to 0^+} h_n(t) = 0$. Next, let

$$r(x) := (x_1^2 + \ldots + x_n^2)^{\frac{1}{2}}$$

and

$$\phi_n(x) := \int_{r(x)}^1 h_n(t)dt$$

for $x \in \overline{B}$. Then $p(x)\Delta\phi_n(x) = -1$, and $\phi_n|_S = 0$.

Theorem 72. *The operator $W : D(W) \to R(W)$ is Ulam stable with constant $\|\phi_n\|$.*

Proof. Let $f \in D(W)$, $\|Wf\| \leq 1$. Then

$$p\Delta f = Wf \in R(W),$$

so that $p\Delta f = pv$, with $v \in C^1(B) \cap C(\overline{B})$. This entails $\Delta f = v$, and Theorem 70 guarantees the existence of $u \in C^2(B) \cap C(\overline{B})$ with $\Delta u = v$ and $u|_S = 0$. Now $p\Delta f = p\Delta u$, i.e.,

$$\|Wu\| = \|Wf\| \leq 1.$$

In particular,

$$p(x)\Delta u(x) \geq -1 = p(x)\Delta\phi_n(x),$$

so that $\Delta(u - \phi_n) \geq 0$. Since $(u - \phi_n)|_S = 0$, we get $u - \phi_n \leq 0$. Similarly we deduce $u + \phi_n \geq 0$, and finally, $-\phi_n \leq u \leq \phi_n$. This means that

$$\|u\| \leq \|\phi_n\|.$$

Now let $g := f - u$. Then

$$Wg = Wf - Wu = p\Delta f - p\Delta u = 0,$$

i.e., $Wg = 0$ and

$$\|f - g\| = \|u\| \leq \|\phi_n\|.$$

So the proof is finished. □

Now let's evaluate the constant $\|\phi_n\|$. Since h_n is nonnegative, we have

$$\|\phi_n\| = \int_0^1 h_n(t)dt,$$

and it is easy to infer that

$$h_n(t) = 2n \sum_{k=0}^\infty \frac{t^{2k+1}}{n + 2k}, \qquad n \geq 1.$$

Therefore,

$$\|\phi_n\| = n \sum_{k=0}^{\infty} \frac{1}{(k+1)(n+2k)}, \qquad n \geq 1.$$

It follows immediately that

$$\|\phi_1\| = \log 4, \qquad \|\phi_2\| = \frac{\pi^2}{6}.$$

Let $n \geq 3$. By using [1, (6.3.16)] with $z = \frac{n}{2} - 1$ we get

$$\psi\left(\frac{n}{2}\right) = -\gamma + (n-2) \sum_{k=0}^{\infty} \frac{1}{(k+1)(n+2k)},$$

where ψ is the Digamma function and $\gamma = -\psi(1)$ is Euler's constant.

So we have

$$\|\phi_n\| = \frac{n}{n-2}\left(\psi\left(\frac{n}{2}\right) + \gamma\right).$$

According to [1, (6.3.4)] and [1, (6.3.6)],

$$\psi\left(n + \frac{1}{2}\right) = 2 \sum_{k=1}^{n} \frac{1}{2k-1} - \log 4 - \gamma,$$

$$\psi(n) = \sum_{k=1}^{n-1} \frac{1}{k} - \gamma.$$

Therefore, we get the following result:

Theorem 73. *An Ulam constant of W is $\|\phi_n\|$, where the following is true:*

$$\|\phi_1\| = \log 4, \qquad \|\phi_2\| = \frac{\pi^2}{6},$$

$$\|\phi_{2m}\| = \frac{m}{m-1} \sum_{k=1}^{m-1} \frac{1}{k}, \qquad m \geq 2,$$

$$\|\phi_{2m+1}\| = \frac{2m+1}{2m-1}\left(2 \sum_{k=1}^{m} \frac{1}{2k-1} - \log 4\right), \qquad m \geq 1.$$

Remark 15. It is known (see, e.g., [9]) that the infimum of the set of Ulam constants for an operator is not necessarily an Ulam constant. It would be interesting to find, for Δ and $p(x)\Delta$, the corresponding infima and to see if they are Ulam constants.

REFERENCES

1. M. Abramowitz, I.A. Stegun (eds.), Handbook of Mathematical Functions with Formulas, Graphs, and Mathematical Tables, National Bureau of Standards, Applied Mathematics Series - 55, Tenth Printing with corrections, Washington, 1972.
2. C. Alsina, R. Ger, On some inequalities and stability results related to the exponential function, J. Inequal. Appl. 2 (1998) 373–380.
3. F. Altomare, M. Campiti, Korovkin-type Approximation Theory and its Applications, W. de Gruyter, Berlin - New York, 1994.
4. F. Altomare, M.C. Montano, V. Leonessa, I. Raşa, Markov Operators, Positive Semigroups and Approximation Processes, De Gruyter Studies in Mathematics, Vol. 61, 2014.
5. S. András, A.R. Mészáros, Ulam-Hyers stability of elliptic partial differential equations in Sobolev spaces, Appl. Math. Comput. 229 (2014) 131–138.
6. D.S. Cîmpean, D. Popa, On the stability of the linear differential equation of higher order with constant coefficients, Appl. Math. Comput. 217 (2010) 4141–4146.
7. D.S. Cîmpean, D. Popa, Hyers-Ulam stability of Euler's equation, Appl. Math. Lett. 24 (2011) 1539–1543.
8. E. Gselmann, Stability properties in some classes of second order partial differential equations, Results Math. 65 (2014) 95–103.
9. O. Hatori, K. Kobayashi, T. Miura, H. Takagi, S.E. Takahasi, On the best constant of Hyers-Ulam stability, J. Nonlinear Convex Anal. 5 (2004) 387–393.
10. B. Hegyi, S.M. Jung, On the stability of Laplace's equation, Appl. Math. Lett. 26 (2013) 549–552.
11. S.M. Jung, Hyers-Ulam stability of linear differential equations of first order, III, J. Math. Anal. Appl. 311 (2005) 139–146.
12. S.M. Jung, Hyers-Ulam stability of linear differential equations of first order, II, Appl. Math. Lett. 19 (2006) 854–858.
13. S.M. Jung, Hyers-Ulam stability of a system of first order linear differential equations with constant coefficients, J. Math. Anal. Appl. 320 (2006) 549–561.
14. S.M. Jung, H. Şevli, Power series method and approximate linear differential equations of second order, Adv. Difference Equ. 2013:76 (2013) 1–9.
15. Y. Li, Y. Shen, Hyers-Ulam stability of linear differential equations of second order, Appl. Math. Lett. 23 (2010) 306–309.
16. N. Lungu, D. Popa, Hyers-Ulam stability of a first order partial differential equation, J. Math. Anal. Appl. 385 (2012) 86–91.
17. N. Lungu, D. Popa, On the Hyers-Ulam stability of a first order partial differential equation, Carpathian J. Math. 28 (2012) 77–82.
18. T. Miura, S. Miyajima, S.E. Takahasi, A characterization of Hyers-Ulam stability of first order linear differential operators, J. Math. Anal. Appl. 286 (2003) 136–146.
19. T. Miura, S. Miyajima, S.E. Takahasi, Hyers-Ulam stability of linear differential operator with constant coefficients, Math. Nachr. 258 (2003) 90–96.
20. M. Obłoza, Hyers stability of the linear differential equation, Rocznik Nauk.-Dydakt. Prace Mat. 13 (1993) 259–270.
21. M. Obłoza, Connections between Hyers and Lyapunov stability of the ordinary differential equations, Rocznik Nauk.-Dydakt. Prace Mat. 14 (1997) 141–146.
22. D. Popa, I. Raşa, On the Hyers-Ulam stability of the linear differential equation, J. Math. Anal. Appl. 381 (2011) 530–537.
23. D. Popa, I. Raşa, Hyers-Ulam stability of the linear differential operator with nonconstant coefficients, Appl. Math. Comput. 219 (2012) 1562–1568.
24. A. Prástaro, Th.M. Rassias, Ulam stability in geometry of PDE's, Nonlinear Funct. Anal. Appl. 8 (2003) 259–278.
25. R. Precup, Linear and Semilinear Partial Differential Equations, De Gruyter, 2013.
26. I. Raşa, Positive operators, Feller semigroups and diffusion equations associated with Altomare projections, Conf. Semin. Mat. Univ. Bari 284 (2002) 1–26.
27. H. Rezaei, S.M. Jung, Th.M. Rassias, Laplace transform and Hyers-Ulam stability of the linear dif-

ferential equations, J. Math. Anal. Appl. 403 (2013) 244–251.

28. I.A. Rus, Remarks on Ulam stability of the operatorial equations, Fixed Point Theory 10 (2009) 305–320.
29. I.A. Rus, Ulam stability of ordinary differential equations, Stud. Univ. Babeş-Bolyai Math. 54 (2009) 125–133.
30. I.A. Rus, N. Lungu, Ulam stability of a nonlinear hyperbolic partial differential equation, Carpathian J. Math. 24 (2008) 403–408.
31. S.E. Takahasi, T. Miura, S. Miyajima, On the Hyers-Ulam stability of the Banach space-valued differential equation $y' = \lambda y$, Bull. Korean Math. Soc. 39 (2002) 309–315.
32. S.E. Takahasi, H. Takagi, T. Miura, S. Miyajima, The Hyers-Ulam stability constants of first order linear differential operators, J. Math. Anal. Appl. 296 (2004) 403–409.
33. G. Wang, M. Zhou, L. Sun, Hyers-Ulam stability of linear differential equations of first order, Appl. Math. Lett. 21 (2008) 1024–1028.

CHAPTER 4

Best constant in Ulam stability

Contents

Abstract

In this section we present some existing results on the best constant in Ulam stability of some classical functional equations and some linear operators in approximation theory.

1. Introduction

Let A, B be two linear spaces with gauges ρ_A and ρ_B, respectively. Let $L : A \to B$ be an operator and $y \in R(L)$, where $R(L)$ denotes the range of L. Recall that the equation

$$Lx = y$$

(or, equivalently, the operator L) is stable in the Ulam sense if there exists $K \geq 0$ with the following property:

for every $\varepsilon > 0$ and every $x \in A$, satisfying $\rho_B(Lx - y) \leq \varepsilon$, there exists $z \in A$ such that $Lz = y$ and

$$\rho_A(x - z) \leq K\varepsilon. \tag{4.1}$$

We call K an Ulam constant for the equation $Lx = y$; see also Chapter 2 and [16]. The infimum of all such constants K satisfying (4.1) is denoted by $K(L)$; generally $K(L)$ does not need to be an Ulam constant for L (see [4]). If $K(L)$ is an Ulam constant for L, then it is called the best Ulam constant, or simply the best constant (denoted by K_L). A similar definition is given in Chapter 2 of this book for linear operators.

Ulam Stability of Operators
http://dx.doi.org/10.1016/B978-0-12-809829-5.50004-0
Copyright © 2018 Elsevier Inc. All rights reserved.

The classical result of Hyers on the stability of the Cauchy functional equation can be interpreted in the sense of the previous definition (see [16]).

Let E and F be Banach spaces and $\|\cdot\|_F$ the norm of F. Let X be the linear space of all mappings on E into F, and define the gauge $\rho_X : X \to [0, +\infty]$ by

$$\rho_X(\varphi) = \sup\{\|\varphi(x)\|_F : x \in E\}, \qquad \varphi \in X.$$

On the linear space $Y = F^{E \times E}$ (as usual, $F^{E \times E}$ denotes the family of all functions mapping $E \times E$ into F) we define a gauge ρ_Y by

$$\rho_Y(\psi) = \sup\{\|\psi(x, y)\|_F : (x, y) \in E \times E\}, \qquad \psi \in Y.$$

Let $L : X \to Y$ be given by

$$L\varphi(x, y) = \varphi(x + y) - \varphi(x) - \varphi(y), \qquad \varphi \in X, \ (x, y) \in E \times E.$$

Then the Hyers result on the stability of Cauchy's functional equation can be reformulated as follows:

For every $\varepsilon > 0$ and every $f \in X$ satisfying $\rho_Y(Lf) \le \varepsilon$ there exists $g \in X$ such that $Lg = 0$ and $\rho_X(f - g) \le \varepsilon$.

The problem of studying the best Ulam constant was first posed in [24]. In the literature there are just a few results on the best constant in the Ulam stability of equations and operators and we mention here the characterization of the stability of linear operators and the representation of their best constants obtained in [4, 5, 17, 26]. The stability of operators has been studied in [10, 11] with some applications to nonlinear analysis; some open problems are also posed there.

In this section we review some existing results (see [20, 21, 22]) and present some new facts concerning the best Ulam constant.

As we mentioned before the infimum of all Ulam constants of an operators is not generally an Ulam constant. For closed operators acting on Hilbert spaces we have the following result:

Theorem 74 ([5]). *Let G and H be Hilbert spaces and $L : D(L) \to H$, where $D(L) \subseteq G$ is the domain of L, a linear closed operator. If L is Ulam stable, then $K(L)$ is the best Ulam constant of L.*

The best Ulam constant for weighted composition operators is obtained in [4]. Let X, Y be compact Hausdorff spaces and $C(X)$, $C(Y)$ be the Banach spaces of all continuous functions defined on X and Y, respectively, endowed with the supremum norm. For $u \in C(Y)$, let

$$S(u) = \{y \in Y : u(y) \ne 0\}.$$

Let $u \in C(Y)$ and $\varphi : Y \to X$ be a function that is continuous on $S(u)$. The function

$uC_\varphi : C(X) \to C(Y)$ defined by

$$(uC_\varphi f)(y) = u(y) \cdot f(\varphi(y)), \qquad f \in C(X), y \in Y,$$

is called a weighted composition operator. Then we have the following result:

Theorem 75 ([4]). *If uC_φ is Ulam stable, then $K(uC_\varphi)$ is the best Ulam constant for uC_φ.*

We present some results on the best Ulam constant for the first-order linear differential operator. Let $h : \mathbb{R} \to \mathbb{C}$ be a continuous function, and consider the linear operator $T_h : C^1(\mathbb{R}, \mathbb{C}) \to C(\mathbb{R}, \mathbb{C})$, given by

$$T_h u = u' + hu, \qquad u \in C^1(\mathbb{R}, \mathbb{C}).$$

Define

$$\|u\|_\infty := \sup_{t \in \mathbb{R}} |u(t)|$$

for all functions $u : \mathbb{R} \to \mathbb{C}$. Clearly $\|\cdot\|_\infty$ is a gauge.

Theorem 76 ([4]). *If T_h is Ulam stable, then $K(T_h)$ is the best Ulam constant for T_h.*

Next, we describe some results obtained for nearisometries. So, let $(E, \|\cdot\|)$ be a real Banach space (possibly a Hilbert space) with $\dim E \geq 1$. For maps $f, g : E \to E$ we set

$$d(f, g) = \sup_{x \in E} \|f(x) - g(x)\|,$$

with the possibility $d(f, g) = \infty$.

We say that a map $f : E \to E$ is a nearisometry if there is a number $\varepsilon \geq 0$ such that

$$\|x - y\| - \varepsilon \leq \|f(x) - f(y)\| \leq \|x - y\| + \varepsilon$$

for all $x, y \in E$. More precisely, such a map is an ε-nearisometry.

D.H. Hyers and S.M. Ulam [9] proved that every surjective ε-nearisometry $f : E \to E$ of a Hilbert space E can be approximated by a surjective isometry $T : E \to E$ such that

$$d(T, f) \leq c\varepsilon.$$

The history of this problem and the best values of c are presented in [6].

2. Best constant for Cauchy, Jensen, and Quadratic functional equations

Let X be a normed space and Y a Banach space over $\mathbb{R}$. We obtain the best Ulam constants for some classical functional equations: Cauchy, Jensen, and Quadratic equations, i.e.,

$$Cf(x, y) := f(x + y) - f(x) - f(y) = 0, \qquad (C)$$

$$Jf(x, y) := f\left(\frac{x + y}{2}\right) - \frac{f(x) + f(y)}{2} = 0, \qquad (J)$$

$$Qf(x, y) := f(x + y) + f(x - y) = 2f(x) + 2f(y) = 0, \qquad (Q)$$

with the unknown $f : X \to Y$.

The first result on the stability of Cauchy's equation, given in the next theorem, was obtained by D.H. Hyers [7].

Theorem 77. *Let $\varepsilon > 0$. Then for every function $f : X \to Y$ satisfying*

$$\|f(x + y) - f(x) - f(y)\| \le \varepsilon, \qquad x, y \in X, \qquad (4.2)$$

there exists a unique additive function $a : X \to Y$ with the property

$$\|f(x) - a(x)\| \le \varepsilon, \qquad x \in X. \qquad (4.3)$$

It is easily seen that (4.3) is optimal. Namely, if $f(x) = \varepsilon u$ for $x \in X$ with a fixed unit vector $u \in Y$, then (4.2) and (4.3) are valid with $a(x) = 0$ for $x \in X$. By the uniqueness of a, this is the best possible estimate.

The first result on Ulam stability of the Jensen equation was given by Z. Kominek [13], but we present below a result given by Y.H. Lee and K.W. Jun [14, Theorem 1.2] (for $p = 0$) on generalized stability of equation (J) (for more information on stability of (J) we refer the reader to [12]).

Theorem 78. *Let $\varepsilon > 0$. Then for every $f : X \to Y$ satisfying the inequality*

$$\left\|f\left(\frac{x + y}{2}\right) - \frac{f(x) + f(y)}{2}\right\| \le \varepsilon, \qquad x, y \in X, \qquad (4.4)$$

there exists a unique additive mapping $a : X \to Y$ such that

$$\|f(x) - a(x) - f(0)\| \le 2\varepsilon, \qquad x \in X. \qquad (4.5)$$

The solutions of equation (Q) are called quadratic functions. It is well known that a function $f : X \to Y$ is quadratic if and only if there exists a unique symmetric and biadditive function $B : X \times X \to Y$ such that

$$f(x) = B(x, x), \qquad x \in X.$$

A first result on Ulam stability of equation (Q) was obtained by F. Skof and generalized later by P.W. Cholewa; for more details see [8, page 45]. The result of Skof and Cholewa is contained in the next theorem.

Theorem 79. *Let $\varepsilon > 0$ and $f : X \to Y$ be a function satisfying*

$$\|f(x + y) + f(x - y) - 2f(x) - 2f(y)\| \le \varepsilon, \qquad x, y \in X. \tag{4.6}$$

Then there exists a unique quadratic function $q : X \to Y$ with the property

$$\|f(x) - q(x)\| \le \frac{\varepsilon}{2}, \qquad x \in X. \tag{4.7}$$

The next theorem has been proved in [22], but we will give a new proof of it without using the uniqueness of the additive/quadratic function in Theorems 77–79.

Theorem 80. *The following relations hold:*
1) $K_C = 1$,
2) $K_J = 2$,
3) $K_Q = \frac{1}{2}$.

Proof. 1) Suppose that Cauchy's equation (C) has an Ulam constant $k < 1$. Let $u \in Y$, $\|u\| = 1$. Consider the function $f : X \to Y$, $f(x) = u$, $x \in X$. The function f satisfies (4.2) with $\varepsilon = 1$, and consequently there exists an additive function $a : X \to Y$ such that

$$\|f(x) - a(x)\| \le k, \qquad x \in X.$$

This entails

$$\|u - a(nx)\| \le k$$

for every $x \in X$ and for every $n \in \mathbb{N}$, where $\mathbb{N}$ is the set of all positive integers.

We get

$$\left\| \frac{u}{n} - a(x) \right\| \le \frac{k}{n}, \qquad x \in X, n \in \mathbb{N}.$$

Letting $n \to \infty$ in the previous relation we get $a(x) = 0$ for $x \in X$, i.e.,

$$1 = \|u\| \le k < 1,$$

which is a contradiction. We conclude that Cauchy's equation cannot have an Ulam constant smaller than 1. Thus, by Theorem 77, Cauchy's equation has the best Ulam constant $K_C = 1$.

2) Suppose that Jensen's equation has an Ulam constant $k < 2$. Let H be a closed hyperplane in X containing the origin, and H_+, H_- be the two open half-spaces determined by H. Let $u \in Y$, $\|u\| = 2$.

Consider the function $f : X \to Y$,

$$f(x) = \begin{cases} u, & x \in H_+, \\ 0, & \text{otherwise.} \end{cases}$$

It is easy to check that

$$\left\| f(\frac{x+y}{2}) - \frac{f(x) + f(y)}{2} \right\| \leq 1, \qquad x, y \in X.$$

So, according to Theorem 78 with $\varepsilon = 1$, there exists an additive function $a : X \to Y$ such that

$$\|f(x) - a(x) - f(0)\| \leq k, \qquad x \in X.$$

Let $x \in H_+$ and $n \in \mathbb{N}$. Then

$$\|f(nx) - a(nx)\| \leq k,$$

which entails

$$\left\| \frac{f(nx)}{n} - a(x) \right\| \leq \frac{k}{n}, \qquad n \in \mathbb{N}.$$

Letting $n \to \infty$ we get $a(x) = 0$, and so

$$2 = \|u\| = \|f(x)\| \leq k < 2,$$

which is a contradiction. By Theorem 78, we conclude that $K_J = 2$.

3) Suppose that the quadratic equation has an Ulam constant $k < \frac{1}{2}$. Let $u \in Y$, $\|u\| = \frac{1}{2}$, and consider the function $f : X \to Y$, $f(x) = u$, $x \in X$. Then (4.6) is satisfied for $\varepsilon = 1$; therefore, there exists a quadratic function $q : X \to Y$ such that

$$\|u - q(x)\| \leq k, \qquad x \in X.$$

On the other hand, there exists a symmetric and biadditive function $B : X \times X \to Y$ such that

$$q(x) = B(x, x), \qquad x \in X.$$

It follows

$$\|u - B(nx, nx)\| \leq k, \qquad x \in X, \, n \in \mathbb{N},$$

which is equivalent to

$$\left\| \frac{u}{n^2} - B(x, x) \right\| \leq \frac{k}{n^2}, \qquad x \in X, \ n \in \mathbb{N}.$$

Letting $n \to \infty$ we get $B(x, x) = 0$, $x \in X$, which entails

$$\frac{1}{2} = \|u\| \leq k < \frac{1}{2},$$

which is a contradiction. By Theorem 79, we conclude that $K_Q = \frac{1}{2}$. $\square$ $\square$

Let us mention yet that, as it can be deduced from the outcomes presented in Chapter 1, we have the following extension of Theorem 77 (see, e.g., [3]):

Theorem 81. *Let X and Y be normed spaces, $c \geq 0$ and $p \neq 1$ be fixed real numbers, and $f : X \to Y$ be a mapping satisfying*

$$\|f(x + y) - f(x) - f(y)\| \leq c(\|x\|^p + \|y\|^p), \qquad x, y \in X \setminus \{0\}. \tag{4.8}$$

If $p \geq 0$ and Y is complete, then there exists a unique function $T : X \to Y$ such that

$$T(x + y) = T(x) + T(y), \qquad x, y \in X, \tag{4.9}$$

and

$$\|f(x) - T(x)\| \leq \frac{c\|x\|^p}{\left|2^{p-1} - 1\right|}, \qquad x \in X \setminus \{0\}. \tag{4.10}$$

If $p < 0$, then f is additive, i.e., it is a solution to (4.9).

Let $c > 0$ and $p \neq 1$ be real numbers, $X_0 := X \setminus \{0\}$, $A := Y^X$, and $B := Y^{X^2}$. Define semigauges $\rho_A : A \to [0, \infty]$ and $\rho_B : B \to [0, \infty]$ by the formulas:

$$\rho_A(f) = \sup_{x \in X_0} \frac{\|f(x)\|}{c\|x\|^p}, \qquad f \in A,$$

$$\rho_B(g) = \sup_{x, y \in X_0} \frac{\|g(x, y)\|}{c(\|x\|^p + \|y\|^p)}, \qquad g \in B.$$

According to Theorem 81, if $p \geq 0$, then $K = |2^{p-1} - 1|^{-1}$ is an Ulam constant for C; next, in the case $X = \mathbb{R}$, it is the smallest possible such constant (cf. [3, Remark 3.7]; see also [2]), which means that it is the best Ulam constant of C. If $p < 0$, then from Theorem 81 we deduce that $K_C = 0$.

3. Best constant for linear operators

Let A, B be normed spaces and $T : A \to B$ a bounded linear operator. Denote by $N(T)$ and $R(T)$ the kernel and range of T, respectively. Next, define the one-to-one operator $\widetilde{T} : A/N(T) \to R(T)$ by

$$\widetilde{T}(f + N(T)) = Tf, \qquad f \in A, \tag{4.11}$$

and let $\widetilde{T}^{-1} : R(T) \to A/N(T)$ be the inverse of $\widetilde{T}$. Then

$$K(T) = \|\widetilde{T}^{-1}\|. \tag{4.12}$$

The main results used in our approach for obtaining, in some concrete cases, the explicit value of $K(T)$ is given by formula (4.12) and a result by D.S. Lubinsky and Z. Ziegler [15], concerning coefficient bounds in the Lorentz representation of a polynomial.

Let $P \in \Pi_n$, where Π_n is the set of all polynomials defined on $[0, 1]$ and of degree, at most, n with real coefficients. Then P has a unique representation of the form

$$P(x) = \sum_{k=0}^{n} c_k x^k (1 - x)^{n-k}, \tag{4.13}$$

where $c_k \in \mathbb{R}$, $k = 0, 1, \ldots, n$. Let T_n denote the Chebyshev polynomial of the first kind. The following representation holds:

$$T_n(2x - 1) = \sum_{k=0}^{n} d_{n,k}(-1)^{n-k} x^k (1 - x)^{n-k} \tag{4.14}$$

where

$$d_{n,k} := \sum_{j=0}^{\min\{k,n-k\}} \binom{n}{2j}\binom{n-2j}{k-j} 4^j, \qquad k = 0, 1, \ldots, n. \tag{4.15}$$

It was proved in [20] that

$$d_{n,k} = \binom{2n}{2k}, \qquad k = 0, 1, \ldots, n. \tag{4.16}$$

Therefore

$$T_n(2x - 1) = \sum_{k=0}^{n} \binom{2n}{2k}(-1)^{n-k} x^k (1 - x)^{n-k}. \tag{4.17}$$

Let

$$\|P\|_\infty = \sup_{x \in [0,1]} |P(x)|.$$

Theorem 82 ([15]). *Let P have representation* (4.13) *for* $x \in [0, 1]$ *and let* $0 \le k \le n$. *Then*

$$|c_k| \le d_{n,k} \cdot \|P\|_\infty$$

with equality if and only if P is a constant multiple of $T_n(2x - 1)$.

3.1. Stancu operators

Let $C[0, 1]$ be the linear space of all continuous functions $f : [0, 1] \to \mathbb{R}$, endowed with the supremum norm denoted by $\| \cdot \|$, and a, b real numbers, $0 \le a \le b$. The Stancu operator [25] $S_n : C[0, 1] \to \Pi_n$ is defined by

$$S_n f(x) = \sum_{k=0}^{n} f\left(\frac{k+a}{n+b}\right)\binom{n}{k}x^k(1 - x)^{n-k}, \ f \in C[0, 1].$$

We have

$$N(S_n) = \left\{ f \in C[0, 1] : \ f\left(\frac{k+a}{n+b}\right) = 0, \ 0 \le k \le n \right\},$$

which is a closed subspace of $C[0, 1]$, and $R(S_n) = \Pi_n$.

The operator $\widetilde{S}_n : C[0, 1]/N(S_n) \to \Pi_n$ is bijective, and $\widetilde{S}_n^{-1} : \Pi_n \to C[0, 1]/N(S_n)$ is bounded since $\dim \Pi_n = n + 1$. So, according to Theorem 54 in Chapter 2, the operator S_n is Ulam stable (see also [20]).

Theorem 83 ([21]). *For* $n \ge 1$ *we have*

$$K(S_n) = \binom{2n}{2\left[\frac{n}{2}\right]} \Big/ \binom{n}{\left[\frac{n}{2}\right]}.$$

Proof. Let $p \in \Pi_n$, $\|p\| \le 1$,

$$p(x) = \sum_{k=0}^{n} c_k(p)x^k(1 - x)^{n-k}.$$

Consider the piecewise affine function $f_p \in C[0, 1]$ defined by

$$f_p(t) = c_0(p), \qquad t \in \left[0, \frac{a}{n+b}\right],$$

$$f_p(t) = c_n(p), \qquad t \in \left[\frac{n+a}{n+b}, 1\right],$$

$$f_p\left(\frac{k+a}{n+b}\right) = c_k(p)/\binom{n}{k}, \qquad 0 \le k \le n.$$

Then $S_n f_p = p$ and

$$\widetilde{S}_n^{-1}(p) = f_p + N(S_n).$$

As usual, the norm of $\widetilde{S}_n^{-1} : \Pi_n \to C[0,1]/N(S_n)$ is defined by

$$\|\widetilde{S}_n^{-1}\| = \sup_{\|p\|\le 1} \|\widetilde{S}_n^{-1}(p)\| = \sup_{\|p\|\le 1} \inf_{h \in N(S_n)} \|f_p + h\|.$$

Clearly

$$\inf_{h \in N(S_n)} \|f_p + h\| = \|f_p\| = \max_{0\le k\le n} |c_k(p)|/\binom{n}{k}.$$

Therefore

$$\|\widetilde{S}^{-1}\| = \sup_{\|p\|\le 1} \max_{0\le k\le n} |c_k(p)|/\binom{n}{k}$$

$$\le \sup_{\|p\|\le 1} \max_{0\le k\le n} d_{n,k} \cdot \|p\|/\binom{n}{k} = \max_{0\le k\le n} d_{n,k}/\binom{n}{k}.$$

On the other hand, let

$$q(x) = T_n(2x-1), \qquad x \in [0,1].$$

Then $\|q\| = 1$ and $|c_k(q)| = d_{n,k}$, $0 \le k \le n$, according to Theorem 82. Consequently

$$\|\widetilde{S}_n^{-1}\| \ge \max_{0\le k\le n} |c_k(q)|/\binom{n}{k} = \max_{0\le k\le n} d_{n,k}/\binom{n}{k}.$$

So,

$$\|\widetilde{S}_n^{-1}\| = \max_{0\le k\le n} \frac{d_{n,k}}{\binom{n}{k}} = \max_{0\le k\le n} \frac{\binom{2n}{2k}}{\binom{n}{k}}.$$

Let

$$a_k = \binom{2n}{2k}/\binom{n}{k}, \qquad 0 \le k \le n.$$

Then

$$\frac{a_{k+1}}{a_k} = \frac{2n-2k-1}{2k+1}, \qquad 0 \le k \le n.$$

The inequality $\dfrac{a_{k+1}}{a_k} \geq 1$ is satisfied if and only if $k \leq \left[\dfrac{n-1}{2}\right]$; therefore

$$\max_{0 \leq k \leq n} a_k = a_{\left[\frac{n-1}{2}\right]+1} = \begin{cases} a_{\left[\frac{n}{2}\right]}, & n \quad \text{even} \\ a_{\left[\frac{n}{2}\right]+1}, & n \quad \text{odd.} \end{cases}$$

Since $a_{\left[\frac{n}{2}\right]+1} = a_{\left[\frac{n}{2}\right]}$ if n is an odd number, we conclude that

$$K(S_n) = \|\widetilde{S}_n^{-1}\| = \binom{2n}{2\left[\frac{n}{2}\right]} \Big/ \binom{n}{\left[\frac{n}{2}\right]}.$$

The theorem is proved.   □

Remark 16. $K(S_n)$ does not depend on a and b. For $a = b = 0$, the Stancu operator reduces to the classical Bernstein operator. Therefore the infimum of the Ulam constants of the Bernstein operator is

$$K(B_n) = \binom{2n}{2\left[\frac{n}{2}\right]} \Big/ \binom{n}{\left[\frac{n}{2}\right]}.$$

3.2. Kantorovich operators

Let

$$X = \{f : [0, 1] \to \mathbb{R} \mid f \text{ is bounded and Riemann integrable}\}$$

be endowed with the supremum norm denoted by $\| \cdot \|$.

The Kantorovich operators [1] are defined by

$$K_n f(x) = (n+1) \sum_{k=0}^{n} \left(\int_{\frac{k}{n+1}}^{\frac{k+1}{n+1}} f(t)dt \right) \binom{n}{k} x^k (1-x)^{n-k}$$

for $f \in X$ and $x \in [0, 1]$. The kernel of K_n is given by

$$N(K_n) = \left\{ f \in X : \int_{\frac{k}{n+1}}^{\frac{k+1}{n+1}} f(t)dt = 0, \ 0 \leq k \leq n \right\}$$

and $N(K_n)$ is a closed subspace of X.

The operators K_n are Ulam stable since their ranges are finite dimensional spaces (see also [20]).

Theorem 84 ([21]). *The following relation holds:*

$$K(K_n) = \binom{2n}{2\left[\frac{n}{2}\right]} \Big/ \binom{n}{\left[\frac{n}{2}\right]}.$$

Proof. Let $p \in \Pi_n$, $\|p\| \leq 1$, and its Lorentz representation

$$p(x) = \sum_{k=0}^{n} c_k(p)x^k(1-x)^{n-k}, \quad x \in [0,1].$$

Consider the piecewise constant function $f_p \in X$ defined by

$$f_p(t) = \begin{cases} \dfrac{c_k(p)}{\dbinom{n}{k}}, & t \in \left[\dfrac{k}{n+1}, \dfrac{k+1}{n+1}\right), \quad 0 \leq k \leq n-1 \\[1em] c_n(p), & t \in \left[\dfrac{n}{n+1}, 1\right]. \end{cases}$$

Then $K_n f_p = p$, i.e.,

$$\widetilde{K}_n^{-1}(p) = f_p + N(K_n).$$

Clearly, we have

$$\|\widetilde{K}_n^{-1}\| = \sup_{\|p\| \leq 1} \|\widetilde{K}_n^{-1}(p)\| = \sup_{\|p\| \leq 1} \inf_{k \in N(K_n)} \|f_p + h\| = \sup_{\|p\| \leq 1} \|f_p\|$$

$$= \sup_{\|p\| \leq 1} \max_{0 \leq k \leq n} \frac{|c_k(p)|}{\dbinom{n}{k}} \leq \sup_{\|p\| \leq 1} \max_{0 \leq k \leq n} \frac{d_{n,k}\|p\|}{\dbinom{n}{k}} = \max_{1 \leq k \leq n} \frac{d_{n,k}}{\dbinom{n}{k}}.$$

Now, let

$$q(x) = T_n(2x-1), \quad x \in [0,1].$$

Then $\|q\| = 1$ and

$$|c_k(q)| = d_{n,k}, \quad 0 \leq k \leq n.$$

Consequently

$$\|\widetilde{K}_n^{-1}\| \geq \max_{0 \leq k \leq n} \frac{|c_k(q)|}{\dbinom{n}{k}} = \max_{0 \leq k \leq n} \frac{d_{n,k}}{\dbinom{n}{k}},$$

and so

$$\|\widetilde{K}_n^{-1}\| = \max_{0 \leq k \leq n} \binom{2n}{2k} \Big/ \binom{n}{k}.$$

As in the proof of Theorem 83 it follows that

$$\|\widetilde{K}_n^{-1}\| = \binom{2n}{2\left[\frac{n}{2}\right]} \Big/ \binom{n}{\left[\frac{n}{2}\right]}.$$

The theorem is proved.   □

Related results can be found in [18, 19].

3.3. An extremal property of $K(B_n)$

We consider a class of generalized positive linear operators defined on $C[0, 1]$ endowed with the supremum norm $\| \cdot \|$. Let $L_n : C[0, 1] \to \Pi_n$ be defined by

$$L_n f(x) := \sum_{k=0}^{n} A_{n,k}(f)\binom{n}{k}x^k(1 - x)^{n-k}, \qquad f \in C[0, 1],$$

where $A_{n,k} : C[0, 1] \to \mathbb{R}$, for $n \in \mathbb{N} \cup \{0\}$ and $0 \leq k \leq n$, are positive linear functionals satisfying $A_{n,k}(1) = 1$ for $0 \leq k \leq n$. Then

$$N(L_n) = \{f \in C[0, 1] : A_{n,k}(f) = 0, \ 0 \leq k \leq n\}$$

and $\widetilde{L}_n : C[0, 1]/N(L_n) \to R(L_n)$ is a bijective operator. Since $R(L_n) \subset \Pi_n$ and $\dim \Pi_n = n + 1$, $\widetilde{L}_n^{-1}$ is bounded. Thus the operator L_n is Ulam stable.

The following result is proved in [21].

Theorem 85 ([21]). *The following inequality holds:*

$$K(L_n) \geq K(B_n).$$

Proof. Let $p \in \Pi_n$,

$$p(x) = \sum_{k=0}^{n} c_k(p)x^k(1 - x)^{n-k}, \qquad c_k(p) \in \mathbb{R}, \ 0 \leq k \leq n.$$

Consider $f \in C[0, 1]$ such that $L_n f = p$. Then

$$A_{n,k}(f) = c_k(p)/\binom{n}{k}, \qquad 0 \leq k \leq n.$$

Now

$$|c_k(p)|/\binom{n}{k} = |A_{n,k}(f)| \leq \|A_{n,k}\| \cdot \|f\| = \|f\|,$$

and thus

$$\max_{0 \leq k \leq n} \frac{|c_k(p)|}{\binom{n}{k}} \leq \|f\|,$$

for all $f \in C[0,1]$ with $L_n f = p$, i.e.,

$$\max_{0 \le k \le n} \frac{|c_k(p)|}{\binom{n}{k}} \le \inf\{\|f\| : L_n f = p\}.$$

On the other hand, let again $q(x) = T_n(2x - 1)$. Then

$$\|\widetilde{L}_n^{-1}\| = \sup_{\|p\| \le 1} \|\widetilde{L}_n^{-1}(p)\| = \sup_{\|p\| \le 1} \inf\{\|f\| : L_n f = p\}$$

$$\ge \sup_{\|p\| \le 1} \max_{0 \le k \le n} \frac{|c_k(p)|}{\binom{n}{k}} \ge \max_{0 \le k \le n} \frac{|c_k(q)|}{\binom{n}{k}}$$

$$= \max_{0 \le k \le n} \frac{d_{n,k}}{\binom{n}{k}} = \|\widetilde{B}_n^{-1}\|.$$

It follows that

$$K(L_n) \ge K(B_n),$$

so the theorem is proved. $\square$

In Theorem 83 and Theorem 84 we have obtained the infimum of Ulam constants for Stancu, Bernstein, and Kantorovich operators. As we mentioned in the introduction, the infimum of all Ulam constants is not generally an Ulam constant. We will prove in what follows that in the case of Stancu, Bernstein, and Kantorovich operators $K(S_n)$, $K(B_n)$, $K(K_n)$ are the best Ulam constants, by showing that their kernels are proximinal sets in the space of $C[0,1]$ endowed with the supremum norm. To this end we will use the following result.

Theorem 86 ([4]). *Suppose that T is an Ulam stable linear operator and its kernel $N(T)$ is a proximinal set. Then $K(T)$ is the best Ulam constant of T.*

Recall that a subset M of the normed space $(A, \| \cdot \|)$ is called a proximinal set if for every $f \in A$ there exists $g \in M$ such that

$$\text{dist}(f, M) = \|f - g\|$$

where

$$\text{dist}(f, M) = \inf\{\|f - h\| : h \in M\}.$$

We present in what follows some results from [23].

Lemma 3. *Let $0 \leq x_0 < x_1 < \ldots < x_n \leq 1$ and*

$$N := \{g \in C[0, 1] : g(x_i) = 0, \ i = 0, 1, \ldots, n\}.$$

Then N is proximinal in $(C[0, 1], \|\cdot\|)$.

Proof. Let $f \in C[0, 1]$ and $m := \max\{|f(x_i)| : \ i = 0, \ldots, n\}$. If $g \in N$, then $\|f - g\| \geq m$, so that

$$\text{dist}(f, N) \geq m. \tag{4.18}$$

If $m = 0$, then $f \in N$; hence $\text{dist}(f, N) = 0 = \|f - f\|$.

It remains to consider the case when $m > 0$. Let $\lambda \in C[0, 1]$ be the piecewise affine function such that

$$\lambda(0) = \lambda(x_0) = \frac{m - f(x_0)}{2m}, \qquad \lambda(x_n) = \lambda(1) = \frac{m - f(x_n)}{2m},$$

$$\lambda(x_i) = \frac{m - f(x_i)}{2m}, \qquad i = 1, \ldots, n - 1.$$

Then

$$0 \leq \lambda(x) \leq 1, \qquad x \in [0, 1].$$

Consider the function

$$g(x) := f(x) + m(2\lambda(x) - 1), \qquad x \in [0, 1].$$

Then $g \in N$ and

$$\text{dist}(f, N) \leq \|f - g\| = m\|2\lambda - 1\| \leq m.$$

Combined with (4.18), this concludes the proof. $\qquad\square$

Lemma 4. *$N(K_n)$ is proximinal in $(X, \|\cdot\|)$.*

Proof. Let

$$I_k := \left[\frac{k}{n+1}, \frac{k+1}{n+1}\right), \qquad k = 0, 1, \ldots, n - 1,$$

and

$$I_n := \left[\frac{n}{n+1}, 1\right].$$

Let $\|h\|_k$ be the supremum norm of a bounded function h defined on I_k for $k = 0, \ldots, n$.

Let $f \in X$ and $g \in N(K_n)$. Then

$$|f(t) - g(t)| \leq \|f - g\|_k, \qquad t \in I_k,$$

and this entails

$$(n + 1)\left|\int_{I_k} f(t)dt\right| \leq \|f - g\|_k, \qquad k = 0, \ldots, n. \qquad (4.19)$$

Set

$$m := \max\left\{(n + 1)\left|\int_{I_k} f(t)dt\right| : k = 0, \ldots, n\right\}.$$

From (4.19) we get $m \leq \|f - g\|$, so

$$m \leq \text{dist}(f, N(K_n)). \qquad (4.20)$$

Now, let $g \in X$ be the function defined by

$$g(t) = f(t) - (n + 1)\int_{I_k} f(s)ds, \qquad t \in I_k, \; k = 0, \ldots, n.$$

Then $g \in N(K_n)$ and $\|f - g\| = m$. Combined with (4.20), this proves the lemma.
□

The results proved in Lemma 3, Lemma 4, and Theorem 86 lead to the following conclusion:

Corollary 7. *The best Ulam constant for the Bernstein, Stancu, and Kantorovich n-th operators is*

$$\binom{2n}{2\left[\frac{n}{2}\right]} \Big/ \binom{n}{\left[\frac{n}{2}\right]}.$$

4. Ulam stability of operators with respect to different norms

For $f \in C[0, 1]$ let $\|f\|_\infty$ be the supremum norm and

$$\|f\|_1 = \int_0^1 |f(x)|dx.$$

Let $n \geq 1$ be given, and

$$N := \left\{g \in C[0, 1] : g\left(\frac{k}{n}\right) = 0, \; k = 0, 1, \ldots, n\right\}.$$

Lemma 5. *N is dense in $(C[0, 1], \|\cdot\|_1)$.*

Proof. Let $f \in C[0, 1]$, $f \neq 0$, $\varepsilon > 0$, and

$$\delta := \min\left\{\frac{1}{2n}, \frac{\varepsilon}{4n\|f\|_\infty}\right\}.$$

Set $I_0 = [0, \delta]$, $I_n = [1 - \delta, 1]$, and

$$I_k = \left[\frac{k}{n} - \delta, \frac{k}{n} + \delta\right], \qquad k = 1, 2, \ldots, n - 1.$$

Let $g \in C[0, 1]$ be the function which is affine on I_0, I_n, and on each of the intervals

$$\left[\frac{k}{n} - \delta, \frac{k}{n}\right], \qquad \left[\frac{k}{n}, \frac{k}{n} + \delta\right], \qquad k = 1, \ldots, n - 1,$$

and

$$g\left(\frac{k}{n}\right) = 0, \qquad k = 0, 1, \ldots, n,$$

$$g(x) = f(x), \qquad x \in [0, 1] \setminus \bigcup_{k=0}^{n} I_k.$$

Then $g \in N$ and $\|g\|_\infty \leq \|f\|_\infty$; hence

$$|f(x) - g(x)| \leq 2\|f\|_\infty, \qquad x \in [0, 1].$$

Let

$$I := \bigcup_{k=0}^{n} I_k.$$

Then

$$\int_0^1 |f(x) - g(x)|dx = \int_I |f(x) - g(x)|dx \leq 4\|f\|_\infty n\delta \leq \varepsilon.$$

Therefore $\|f - g\|_1 \leq \varepsilon$, and this concludes the proof. $\square$

Theorem 87. *Let $K > 0$. The Bernstein operator*

$$B_n : (C[0, 1], \|\cdot\|_1) \to (C[0, 1], \|\cdot\|_1)$$

is Ulam stable with an Ulam constant K.

Proof. Let $f \in C[0, 1]$ with $\|B_n f\|_1 \leq 1$. According to Lemma 5, there exists $g \in N = \ker B_n$ such that $\|f - g\|_1 \leq K$. This means that K is an Ulam constant for B_n in view of Remark 8 in Chapter 2. (Let us remark that the assumption $\|B_n f\|_1 \leq 1$ has not been used in the proof!). $\square$

Remark 17. Theorem 87 provides an example of an operator for which the infimum of the Ulam constants is 0, and this infimum is not an Ulam constant. Another example of an operator T for which the infimum K_T is not an Ulam constant can be found in [4].

In what follows, for $f \in C_b[0, +\infty)$ let $\|f\|_\infty$ be the supremum norm; consider also the generalized norm

$$\|f\|_1 := \int_0^\infty |f(x)| dx.$$

Let $n \geq 1$ be given, and

$$M := \left\{ g \in C_b[0, +\infty), \ g\left(\frac{k}{n}\right) = 0, \ k = 0, 1, \ldots \right\}.$$

Lemma 6. *M is dense in* $(C_b[0, +\infty), \|\cdot\|_1)$, *i.e., for each* $\varepsilon > 0$ *and for each* $f \in C_b[0, +\infty)$ *there exists* $g \in M$ *with* $\|f - g\|_1 \leq \varepsilon$.

Proof. Let $f \in C_b[0, +\infty)$, $f \neq 0$, and $\varepsilon > 0$. Let

$$0 < \delta_k \leq \frac{1}{2n}, \qquad k \geq 0,$$

be such that

$$\delta_0 + 2 \sum_{k=1}^\infty \delta_k \leq \frac{\varepsilon}{2\|f\|_\infty}.$$

Consider the intervals $I_0 = [0, \delta_0]$ and

$$I_k = \left[\frac{k}{n} - \delta_k, \frac{k}{n} + \delta_k\right], \qquad k \geq 1.$$

Let $g \in C_b[0, +\infty)$ be the function which is affine on I_0 and on each of the intervals

$$\left[\frac{k}{n} - \delta_k, \frac{k}{n} + \delta_k\right], \qquad k \geq 1,$$

and

$$g\left(\frac{k}{n}\right) = 0, \qquad k \geq 0,$$

$$g(x) = f(x), \qquad x \in [0, +\infty) \setminus \bigcup_{k=0}^\infty I_k.$$

Then $g \in M$ and $\|g\|_\infty \leq \|f\|_\infty$; hence

$$|f(x) - g(x)| \leq 2\|f\|_\infty, \qquad x \in [0, +\infty).$$

Let

$$I := \bigcup_{k=0}^{\infty} I_k.$$

Then

$$\int_0^\infty |f(x) - g(x)| dx = \int_I |f(x) - g(x)| dx$$

$$\leq 2\|f\|_\infty \left(\delta_0 + 2 \sum_{k=1}^{\infty} \delta_k \right) \leq \varepsilon.$$

Therefore $\|f - g\|_1 \leq \varepsilon$, which shows that M is dense in $(C_b[0, +\infty), \|\cdot\|_1)$. $\square$

Theorem 88. *Let $K > 0$ and $L_n : (C_b([0, +\infty), \|\cdot\|_1) \to (C_b[0, +\infty), \|\cdot\|_1)$ be the Szász-Mirakjan operator defined by*

$$L_n f(x) := e^{-nx} \sum_{i=0}^{\infty} f\left(\frac{i}{n}\right) \frac{n^i}{i!} x^i, \ f \in C_b[0, +\infty), \qquad x \geq 0.$$

Then L_n is Ulam stable with Ulam constant K.

Proof. Similar to that of Theorem 87. $\square$

Remark 18. $L_n : (C_b[0, +\infty), \|\cdot\|_\infty) \to (C_b[0, +\infty), \|\cdot\|_\infty)$ is not Ulam stable; see [20].

REFERENCES

1. F. Altomare, M. Campiti, Korovkin-type Approximation Theory and its Applications, W. de Gruyter, Berlin - New York, 1994.
2. J. Brzdęk, A note on stability of additive mappings, In: Stability of Mappings of Hyers-Ulam Type (Th.M. Rassias, J. Tabor eds.), pp. 19–22, Hadronic Press, Palm Harbor, FL, 1994.
3. J. Brzdęk, W. Fechner, M.S. Moslehian, J. Sikorska, Recent developments of the conditional stability of the homomorphism equation, Banach J. Math. Anal. 9 (2015) 278–326.
4. O. Hatori, K. Kobayashi, T. Miura, H. Takagi, S.E. Takahasi, On the best constant of Hyers-Ulam stability, J. Nonlinear Convex Anal. 5 (2004) 387–393.
5. G. Hirasawa, T. Miura, Hyers-Ulam stability of a closed operator in a Hilbert space, Bull. Korean Math. Soc. 43 (2006) 107–117.
6. T. Huuskonen, J. Väisälä, Hyers-Ulam constants of Hilbert spaces, Stud. Math. 153 (2002) 31–40.
7. D.H. Hyers, On the stability of the linear functional equation, Proc. Nat. Acad. Sci. U.S.A. 27 (1941) 222–224.
8. D.H. Hyers, G. Isac, Th.M. Rassias, Stability of Functional Equations in Several Variables, Birkhäuser Boston, Inc., Boston, MA, 1998.
9. D.H. Hyers, S.M. Ulam, On approximate isometries, Bull. Amer. Math. Soc. 51 (1945) 288–292.

10. G. Isac, Th.M. Rassias, On the Hyers-Ulam stability of ψ-additive mappings, J. Approx. Theory 72 (1993) 131–137.
11. G. Isac, Th.M. Rassias, Stability of ψ-additive mappings: applications to nonlinear analysis, Int. J. Math. Math. Sci. 19 (1996) 219–228.
12. S.M. Jung, Hyers-Ulam-Rassias stability of Jensen's equation and its application, Proc. Amer. Math. Soc. 126 (1998) 3137–3143.
13. Z. Kominek, On a local stability of the Jensen functional equation, Demonstratio Math. 22 (1989) 499–507.
14. Y.H. Lee, K.W. Jun, A generalization of the Hyers-Ulam-Rassias stability of the Pexider equation, J. Math. Anal. Appl. 246 (2000) 627–638.
15. D.S. Lubinsky, Z. Ziegler, Coefficient bounds in the Lorentz representation of a polynomial, Canad. Math. Bull. 33 (1990) 197–206.
16. T. Miura, S. Miyajima, S.E. Takahasi, Hyers-Ulam stability of linear differential operator with constant coefficients, Math. Nachr. 258 (2003) 90–96.
17. T. Miura, S. Miyajima, S.E. Takahasi, A characterization of Hyers-Ulam stability of first order linear differential operators, J. Math. Anal. Appl. 286 (2003) 136–146.
18. M. Mursaleen, K.J. Ansari, On the stability of some positive linear operators from approximation theory, Bull. Math. Sci. 5 (2015) 147–157.
19. M. Mursaleen, K.J. Ansari, K. Asif, Stability of some positive linear operators on compact disk, Acta Math. Sci. 35 (2015) 1492–1500.
20. D. Popa, I. Raşa, On the stability of some classical operators from approximation theory, Expo. Math. 31 (2013) 205–214.
21. D. Popa, I. Raşa, On the best constant in Hyers-Ulam stability of some positive linear operators, J. Math. Anal. Appl. 412 (2014) 103–108.
22. D. Popa, I. Raşa, Best constant in Hyers-Ulam stability of some functional equations, Carpathian J. Math. 30 (2014) 383–386.
23. D. Popa, I. Raşa, Best constant in stability of some positive linear operators, Aequationes Math. 90 (2016) 719–726.
24. Th.M. Rassias, J. Tabor, What is left of Hyers-Ulam stability?, J. Natur. Geom. 1 (1992) 65–69.
25. D.D. Stancu, Asupra unei generalizări a polinoamelor lui Bernstein, Studia Univ. Babeş-Bolyai 14 (1969) 31–45.
26. H. Takagi, T. Miura, S.E. Takahasi, Essential norms and stability constants of weighted composition operators on $C(X)$, Bull. Korean Math. Soc. 40 (2003) 583–591.

CHAPTER 5

Ulam stability of operators of polynomial form

Contents

Abstract

The linear difference and functional equations in a single variable have been deeply studied in the literature. They are examples of equations defined by operators of polynomial form. The results on stability of such equations are obtained by the iterative method and fixed point method. There are numerous interesting outcomes in this area that have been proved in recent years. A complete characterization of Ulam's stability is available for the linear difference and functional equations of higher order with constant coefficients.

1. Introduction

Throughout this section $\mathbb{R}_+ := [0, \infty)$, $\mathbb{K} \in \{\mathbb{R}, \mathbb{C}\}$, X stands for a normed space over $\mathbb{K}$, and S is a nonempty set (unless explicitly stated otherwise).

The following definition describes the main ideas of the kind of stability that we refer to in this chapter (as usual, B^A denotes the family of all functions mapping a nonempty set A into a nonempty set B).

Definition 8. Let $n \in \mathbb{N}$, A be a nonempty set, (X, d) be a metric space, $\mathcal{E} \subset C \subset \mathbb{R}_+^{A^n}$ be nonempty, $\mathcal{T}$ be an operator (not necessarily linear) mapping C into $\mathbb{R}_+^A$, and $\mathcal{F}_1, \mathcal{F}_2$

http://dx.doi.org/10.1016/B978-0-12-809829-5.50005-2
Copyright © 2018 Elsevier Inc. All rights reserved.

be operators (not necessarily linear) mapping nonempty $\mathcal{D} \subset X^A$ into X^{A^n}. We say that the equation

$$\mathcal{F}_1 \varphi(x_1, \ldots, x_n) = \mathcal{F}_2 \varphi(x_1, \ldots, x_n) \tag{5.1}$$

is $(\mathcal{E}, \mathcal{T})$ – stable provided, for every $\varepsilon \in \mathcal{E}$ and $\varphi_0 \in \mathcal{D}$ with

$$d(\mathcal{F}_1 \varphi_0(x_1, \ldots, x_n), \mathcal{F}_2 \varphi_0(x_1, \ldots, x_n)) \leq \varepsilon(x_1, \ldots, x_n), \qquad x_1, \ldots, x_n \in A, \tag{5.2}$$

there exists a solution $\varphi \in \mathcal{D}$ of equation (5.1) such that

$$d(\varphi(x), \varphi_0(x)) \leq \mathcal{T}\varepsilon(x), \qquad x \in A. \tag{5.3}$$

Roughly speaking, $(\mathcal{E}, \mathcal{T})$ – stability of equation (5.1) means that every approximate (in the sense of (5.2)) solution of (5.1) is always close (in the sense of (5.3)) to an exact solution to (5.1).

In the particular case when $\mathcal{E}$ contains only all constant functions, the $(\mathcal{E}, \mathcal{T})$ – stability is usually called the *Hyers-Ulam stability* or *stability of the Hyers-Ulam type*.

Clearly, X^S (the family of all the functions mapping S into X) is a linear space over $\mathbb{K}$, with the usual linear structure, defined as follows:

$$(f + g)(x) = f(x) + g(x), \qquad f, g \in X^S, x \in S,$$

$$(\alpha f)(x) = \alpha f(x), \qquad f \in X^S, x \in S, \alpha \in \mathbb{K}.$$

For simplicity, given $f \in X^S$, we write

$$\|f\|^* := \sup_{x \in S} \|f(x)\|,$$

and we define the function $\|f\| : S \to \mathbb{R}$ by $\|f\|(x) = \|f(x)\|$ for each $x \in S$.

In what follows, $\mathcal{U}$ denotes a linear subspace of X^S, $F \in X^S$ is a fixed function, and $\mathcal{L} : \mathcal{U} \to X^S$ is linear. Moreover, we always assume that $\mathcal{U}$ is nontrivial, i.e., it contains nonzero elements.

Let $m \in \mathbb{N}$ be fixed (in general we assume that $m > 1$, unless explicitly stated otherwise), $a_0, \ldots, a_{m-1} \in \mathbb{K}$ and $P_m : \mathbb{C} \to \mathbb{C}$ be the polynomial given by

$$P_m(z) := z^m + \sum_{j=0}^{m-1} a_j z^j, \qquad z \in \mathbb{C}. \tag{5.4}$$

Write

$$\mathcal{U}_m := \{ f \in \mathcal{U} : \mathcal{L}^i f \in \mathcal{U}, \ i = 1, \ldots, m - 1 \}$$

and define $P_m(\mathcal{L}) : \mathcal{U}_m \to X^S$ by

$$P_m(\mathcal{L}) := \mathcal{L}^m + \sum_{j=0}^{m-1} a_j \mathcal{L}^j,$$

where $\mathcal{L}^0 := \mathcal{I}$ is the identity operator (i.e., $\mathcal{I}f = f$ for $f \in X^S$) and $\mathcal{L}^k := \mathcal{L} \circ \mathcal{L}^{k-1}$ for any $k \in \mathbb{N}$. Note that $\mathcal{U}_m$ is a linear subspace of X^S and $P_m(\mathcal{L})$ is linear, because $\mathcal{L}$ is linear.

In this chapter we present some stability results for the operator

$$\mathcal{P}_m^F := P_m(\mathcal{L}) - F$$

(i.e., $\mathcal{P}_m^F(\varphi) := P_m(\mathcal{L})(\varphi) - F$ for $\varphi \in \mathcal{U}_m$). Thus we provide some general methods for investigation of stability of various linear equations of higher orders, of the form

$$(\mathcal{L}^m \varphi)(x) + \sum_{j=0}^{m-1} a_j (\mathcal{L}^j \varphi)(x) = F(x). \tag{5.5}$$

The following three equations (under suitable assumptions on the unknown function φ) are simple natural particular cases of (5.5):

a) the well-known linear functional equation

$$\varphi(f^m(z)) + a_{m-1}\varphi(f^{m-1}(z)) + \cdots + a_1\varphi(f(z)) + a_0\varphi(z) = F(z), \tag{5.6}$$

b) the linear difference equation

$$\varphi(n + m) + a_{m-1}\varphi(n + m - 1) + \cdots + a_1\varphi(n + 1) + a_0\varphi(n) = F(n), \tag{5.7}$$

c) and the linear differential equation

$$\varphi^{(m)}(z) + a_{m-1}\varphi^{(m-1)}(z) + \cdots + a_1\varphi'(z) + a_0\varphi(z) = F(z). \tag{5.8}$$

Equation (5.6) is one of the most important functional equation. Many results on its continuous solutions, analytic solutions, and integrable solutions can be found in [23, 24] and the references therein.

Equation (5.7) is a discrete case of equation (5.6). Its Hyers-Ulam stability has been discussed, in particular, in [6, 9, 10, 28, 29].

Equation (5.8) is a basic ordinary differential equation. There are numerous results on existence, uniqueness, and Liapunov stability of its general solution. Its stability has been already discussed in previous chapters.

At the end of this chapter we present some particular stability results for equations (5.6)–(5.8), obtained with those general methods.

In the sequel, $p_1, \ldots, p_m \in \mathbb{C}$ denote the roots of the equation

$$P_m(z) = 0. \tag{5.9}$$

Remark 19. If $m > 1$, then

$$P_m(z) = z^m + \sum_{j=0}^{m-1} a_j z^j = (z - p_m)\left(z^{m-1} + \sum_{j=0}^{m-2} b_j z^j\right), \qquad z \in \mathbb{C},$$

for some unique $b_0, \ldots, b_{m-2} \in \mathbb{C}$. It is easily seen that

$$a_{m-1} = -p_m + b_{m-2}, \quad a_0 = -p_m b_0$$

and, in the case $m > 3$,

$$a_j = -p_m b_j + b_{j-1}$$

for $j = 1, \ldots, m - 2$. Moreover, $p_1, \ldots, p_{m-1}$ are roots of the equation

$$z^{m-1} + \sum_{j=0}^{m-2} b_j z^j = 0.$$

Observe yet that, if $p_m, a_0, \ldots, a_{m-1} \in \mathbb{R}$, then we have $b_0, \ldots, b_{m-2} \in \mathbb{R}$.

2. Auxiliary results

Let us introduce one more (technical) definition of stability that will be useful for the next two auxiliary propositions.

Let $\delta, L \in \mathbb{R}_+^S$, $\mathcal{V}_0 \subset \mathcal{V} \subset X^S$ be nonempty, $\mathcal{T} : \mathcal{V} \to X^S$, and $F : S \to X$ be a given function. We say that the equation

$$\mathcal{T}\varphi = F \tag{5.10}$$

is (δ, L)-stable in $\mathcal{V}_0$, provided, for each $\gamma \in \mathcal{V}_0$ with

$$\left\|\mathcal{T}\gamma(x) - F(x)\right\| \leq \delta(x), \qquad x \in S, \tag{5.11}$$

there is a solution $\varphi \in \mathcal{V}$ of (5.10) with

$$\|\gamma(x) - \varphi(x)\| \leq L(x), \qquad x \in S. \tag{5.12}$$

In the case $\mathcal{V}_0 = \mathcal{V}$ we omit the part "in $\mathcal{V}_0$". Moreover, if the function φ is unique, then we say that the equation is stable with uniqueness.

Now we prove four propositions that are very useful tools and concern stability of the linear equations of the general form (5.10) and (5.5), respectively. The first one corresponds to the results in [2].

Proposition 4. *Let $\mathcal{V}_0 \subset \mathcal{V}$ be two linear subspaces of X^S, $\mathcal{T} : \mathcal{V} \to X^S$ be a linear operator, $\delta, L \in \mathbb{R}_+^S$ and $F_1, F_2 : S \to X$. Suppose that the equation*

$$\mathcal{T}f = F_2 - F_1 \tag{5.13}$$

admits a solution $f_0 \in \mathcal{V}_0$. Then the functional equation

$$\mathcal{T}f = F_2 \tag{5.14}$$

is (δ, L)–stable in $\mathcal{V}_0$ (with uniqueness) if and only if the equation

$$\mathcal{T}f = F_1 \tag{5.15}$$

is (δ, L)–stable in $\mathcal{V}_0$ (with uniqueness).

Proof. Assume first that equation (5.14) is (δ, L)–stable. Let $g \in \mathcal{V}_0$ satisfy the condition

$$\|\mathcal{T}g - F_1\| \le \delta. \tag{5.16}$$

Write $g_0 := g + f_0$. Then $g_0 \in \mathcal{V}_0$ and

$$\|\mathcal{T}g_0 - F_2\| = \|\mathcal{T}g - F_1\| \le \delta.$$

Hence, there exists a solution $h_0 \in \mathcal{V}$ of equation (5.14) such that

$$\|g_0 - h_0\| \le L.$$

Clearly, $h := h_0 - f_0 \in \mathcal{V}$ is a solution to (5.15) and

$$\|g - h\| = \|g_0 - h_0\| \le L.$$

The proof of the necessary condition is analogous.

Assume now that equation (5.14) is (δ, L)–stable with uniqueness. Let $g \in \mathcal{V}_0$ satisfy (5.16) and $h, h' \in \mathcal{V}$ be solutions to (5.15) such that

$$\|g - h\| \le L, \qquad \|g - h'\| \le L.$$

Write $g_0 := g + f_0$, $h_0 := h + f_0$ and $h'_0 := h' + f_0$. Then

$$\|\mathcal{T}g_0 - F_2\| \le \delta,$$

h_0 and h'_0 are solutions to (5.14) and

$$\|g_0 - h_0\| \le L, \qquad \|g_0 - h'_0\| \le L.$$

Consequently, $h_0 = h'_0$, whence $h = h'$.

The proof of the converse implication is analogous.   □

Remark 20. The assumption of Proposition 4 that equation (5.13) admits a solution $f_0 \in \mathcal{V}_0$ seems to be quite natural, because if (5.13) does not possess any such solution, then we can consider it to be not stable in a "trivial" way (but we refer to [2] for further comments on that issue).

For the next two propositions we need to recall the notion of complexification of a normed space.

Namely, in the case $\mathbb{K} = \mathbb{R}$, $\mathfrak{C}(X)$ will always denote the set X^2 endowed with a linear structure and the Taylor norm $\|\cdot\|_T$, given by the following:

$$(x, y) + (z, w) := (x + z, y + w),$$

$$(\alpha + i\beta)(x, y) := (\alpha x - \beta y, \beta x + \alpha y),$$

$$\|(x, y)\|_T := \sup_{0 \le \theta \le 2\pi} \|(\cos \theta)x + (\sin \theta)y\|$$

for $x, y, z, w \in X$, $\alpha, \beta \in \mathbb{R}$. It is easy to check that $\mathfrak{C}(X)$ is a complex normed space and

$$\max\{\|x\|, \|y\|\} \le \|(x, y)\|_T \le \|x\| + \|y\|, \qquad x, y \in X. \tag{5.17}$$

Moreover, $\mathfrak{C}(X)$ is a Banach space when X is a Banach space (cf., e.g., [17, p. 39] or [22, 1.9.6, p. 66]).

We define $\pi_1, \pi_2 : X^2 \to X$ by the following:

$$\pi_i(x_1, x_2) := x_i, \qquad x_1, x_2 \in X, i = 1, 2. \tag{5.18}$$

Let

$$\widehat{\mathcal{U}} := \{\mu : S \to X^2 : \pi_i \circ \mu \in \mathcal{U}, i = 1, 2\} \tag{5.19}$$

and $\widehat{\mathcal{L}} : \widehat{\mathcal{U}} \to \mathfrak{C}(X)^S$ be given by

$$\widehat{\mathcal{L}}\mu = (\mathcal{L}(\pi_1 \circ \mu), \mathcal{L}(\pi_2 \circ \mu)), \qquad \mu \in \widehat{\mathcal{U}}. \tag{5.20}$$

It is easily seen that actually

$$\widehat{\mathcal{U}} = \mathcal{U} \times \mathcal{U}.$$

We write

$$\|\mathcal{L}\| := \inf\{\lambda \in \mathbb{R}_+ : \|\mathcal{L}f - \mathcal{L}g\|^* \le \lambda\|f - g\|^* \text{ for all } f, g \in \mathcal{U}\},$$

$$\|\widehat{\mathcal{L}}\|_T := \inf\{\lambda \in \mathbb{R}_+ : \|\widehat{\mathcal{L}}\mu - \widehat{\mathcal{L}}v\|_T^* \le \lambda\|\mu - v\|_T^* \text{ for all } \mu, v \in \widehat{\mathcal{U}}\}.$$

We have the following simple observations (as usual, $\mathfrak{R}(w)$ and $\mathfrak{I}(w)$ denote the real and imaginary parts of a complex number w).

Proposition 5. *Assume that $\mathbb{K} = \mathbb{R}$. Then $\widehat{\mathcal{U}}$ is a linear subspace of $\mathfrak{C}(X)$ over $\mathbb{C}$ and $\widehat{\mathcal{L}}$ is $\mathbb{C}$-linear. Moreover,*

$$\|\mathcal{L}\| \le \|\widehat{\mathcal{L}}\|_T \le \sqrt{2}\|\mathcal{L}\|.$$

Proof. Note that, for every $\mu \in \widehat{\mathcal{U}}$ and $w \in \mathbb{C}$,

$$w\mu = (\mathfrak{R}(w)\pi_1 \circ \mu - \mathfrak{I}(w)\pi_2 \circ \mu, \ \mathfrak{R}(w)\pi_2 \circ \mu + \mathfrak{I}(w)\pi_1 \circ \mu),$$

whence $w\mu \in \widehat{\mathcal{U}}$ (because $\mathcal{U}$ is a linear subspace of X). Moreover, it is easily seen that

$$\mu + \nu \in \widehat{\mathcal{U}}, \qquad \mu, \nu \in \widehat{\mathcal{U}}.$$

Thus we have shown that $\widehat{\mathcal{U}}$ is a linear subspace of $\mathfrak{C}(X)$ (over $\mathbb{C}$).

Next, fix $\mu \in \widehat{\mathcal{U}}$. Clearly, $\mu_1 := \pi_1 \circ \mu, \mu_2 := \pi_2 \circ \mu \in \mathcal{U}$, and, for each $w \in \mathbb{C}$,

$$\mathcal{L}(\mathfrak{R}(w)\mu_1 - \mathfrak{I}(w)\mu_2) = \mathfrak{R}(w)\mathcal{L}\mu_1 - \mathfrak{I}(w)\mathcal{L}\mu_2,$$

$$\mathcal{L}(\mathfrak{I}(w)\mu_1 + \mathfrak{R}(w)\mu_2) = \mathfrak{I}(w)\mathcal{L}\mu_1 + \mathfrak{R}(w)\mathcal{L}\mu_2,$$

whence

$$\begin{aligned}
\widehat{\mathcal{L}}(w\mu) &= \widehat{\mathcal{L}}((\mathfrak{R}(w)\mu_1 - \mathfrak{I}(w)\mu_2, \ \mathfrak{R}(w)\mu_2 + \mathfrak{I}(w)\mu_1)) \\
&= (\mathcal{L}(\mathfrak{R}(w)\mu_1 - \mathfrak{I}(w)\mu_2), \ \mathcal{L}(\mathfrak{R}(w)\mu_2 + \mathfrak{I}(w)\mu_1)) \\
&= (\mathfrak{R}(w)\mathcal{L}\mu_1 - \mathfrak{I}(w)\mathcal{L}\mu_2, \ \mathfrak{I}(w)\mathcal{L}\mu_1 + \mathfrak{R}(w)\mathcal{L}\mu_2) \\
&= w(\mathcal{L}\mu_1, \ \mathcal{L}\mu_2) = w\widehat{\mathcal{L}}\mu.
\end{aligned}$$

Moreover, since $\mathcal{L}$ is additive, so is $\widehat{\mathcal{L}}$. Consequently, $\widehat{\mathcal{L}}$ is $\mathbb{C}$-linear.

Clearly, by the definition of the Taylor norm and (5.17),

$$\begin{aligned}
\|\mathcal{L}f\|^* &= \left\|(\mathcal{L}f, 0)\right\|_T^* = \left\|\widehat{\mathcal{L}}(f, 0)\right\|_T^* \\
&\leq \left\|\widehat{\mathcal{L}}\right\|_T \|(f, 0)\|_T^* = \left\|\widehat{\mathcal{L}}\right\|_T \|f\|^*, \qquad f \in \mathcal{U},
\end{aligned}$$

whence

$$\|\mathcal{L}\| \leq \|\widehat{\mathcal{L}}\|_T.$$

Further, for each $\mu \in \widehat{\mathcal{U}}$ we have

$$\begin{aligned}
\|\widehat{\mathcal{L}}\mu\|_T^* &= \left\|(\mathcal{L}(\pi_1 \circ \mu), \mathcal{L}(\pi_2 \circ \mu))\right\|_T^* \\
&\leq \sup_{0 \leq \theta \leq 2\pi} (|\cos\theta| \, \|\mathcal{L}(\pi_1 \circ \mu)\|^* + |\sin\theta| \, \|\mathcal{L}(\pi_2 \circ \mu)\|^*) \\
&\leq \|\mathcal{L}\| \sup_{0 \leq \theta \leq 2\pi} (|\cos\theta| \, \|\pi_1 \circ \mu\|^* + |\sin\theta| \, \|\pi_2 \circ \mu\|^*) \\
&\leq \|\mathcal{L}\| \sup_{0 \leq \theta \leq 2\pi} (|\cos\theta| + |\sin\theta|) \max \{\|\pi_1 \circ \mu\|^*, \|\pi_2 \circ \mu\|^*\} \\
&\leq \sqrt{2} \, \|\mathcal{L}\| \, \|\mu\|_T^*.
\end{aligned}$$

$\square$

Proposition 6. *Let $K = \mathbb{R}$, $\mathcal{V}_0 \subset \mathcal{V}$ be two linear subspaces of X^S, $\mathcal{T} : \mathcal{V} \to X^S$ be linear, $\delta, L \in \mathbb{R}_+^S$, $F \in X^S$, $F_0(t) := (F(t), 0)$ for $t \in S$, and $\widehat{\mathcal{T}} : \mathcal{V} \times \mathcal{V} \to \mathbb{C}(X)^S$ be defined by:*

$$\widehat{\mathcal{T}} \mu = (\mathcal{T}(\pi_1 \circ \mu), \mathcal{T}(\pi_2 \circ \mu)), \qquad \mu \in \widehat{\mathcal{V}} := \mathcal{V} \times \mathcal{V}. \tag{5.21}$$

Assume that the equation

$$\widehat{\mathcal{T}} \varphi = F_0 \tag{5.22}$$

is (δ, L)-stable in $\mathcal{V}_0 \times \mathcal{V}_0$. Then the equation

$$\mathcal{T} f = F \tag{5.23}$$

is (δ, L)-stable in $\mathcal{V}_0$.

Proof. Let $f \in \mathcal{V}_0$ satisfy

$$\|(\mathcal{T} f)(x) - F(x)\| \leq \delta(x), \qquad x \in S.$$

Write $\varphi_0(x) = (f(x), 0)$ for $x \in S$. Then

$$\left\|(\widehat{\mathcal{T}} \varphi_0)(x) - F_0(x)\right\|_T \leq \delta(x), \qquad x \in S.$$

So, there is a solution $\varphi = (\varphi_1, \varphi_2) \in \widehat{\mathcal{V}}$ of (5.22) such that

$$\|\varphi_0(x) - \varphi(x)\|_T \leq L(x), \qquad x \in S. \tag{5.24}$$

Let $\varphi_i := \pi_i \circ \varphi$ for $i = 1, 2$. Clearly,

$$(\mathcal{T} \varphi_1(z), \mathcal{T} \varphi_2(z)) = \widehat{\mathcal{T}} \varphi(z) = F_0(z) = (F(z), 0), \qquad z \in S.$$

Now, it is easily seen that φ_1 is a solution of (5.23). Moreover, by (5.17) and (5.24),

$$\|f(x) - \varphi_1(x)\| \leq L(x), \qquad x \in S.$$

$\square$

The next auxiliary proposition shows possible applications of the complexification (several other such examples are provided in the further parts of the book). We need in it the following simple observation. Namely, if $\mathcal{L} : \mathcal{U} \to \mathcal{U}$ is bijective, then

$$\|g\|^* \leq \|\mathcal{L}^{-1}(\mathcal{L}g)\|^* \leq \|\mathcal{L}^{-1}\| \|\mathcal{L}\| \|g\|^*, \qquad g \in \mathcal{U},$$

which implies that $1 \leq \|\mathcal{L}^{-1}\| \|\mathcal{L}\|$, and consequently, in the case $\|\mathcal{L}^{-1}\| < \infty$,

$$\frac{1}{\|\mathcal{L}^{-1}\|} \leq \|\mathcal{L}\|. \tag{5.25}$$

Proposition 7. *Assume that $\mathcal{L}(\mathcal{U}) \subset \mathcal{U}$ and one of the following two conditions is valid.*

(A) $|p_i| > \rho$ for $i = 1, \ldots, m$;

(B) $\mathcal{L} : \mathcal{U} \to \mathcal{U}$ is bijective, $\|\mathcal{L}\| < \infty$ and $|p_i| \notin [\rho_0, \rho]$ for $i = 1, \ldots, m$,

   where

$$\rho := \begin{cases} \|\mathcal{L}\|, & \text{if } p_i \in \mathbb{K} \text{ for } i = 1, \ldots, m; \\ \|\widehat{\mathcal{L}}\|_T, & \text{otherwise}, \end{cases}$$

$$\rho_0 := \begin{cases} \|\mathcal{L}^{-1}\|^{-1}, & \text{if } p_i \in \mathbb{K} \text{ for } i = 1, \ldots, m; \\ \|\widehat{\mathcal{L}^{-1}}\|_T^{-1}, & \text{otherwise}. \end{cases}$$

Then, for every $F, \gamma \in X^S$, there is at most one solution $\varphi \in \mathcal{U}$ of (5.5) such that

$$\|\gamma - \varphi\|^* < \infty.$$

Proof. Notice that if $\gamma \in X^S$ and $\varphi_1, \varphi_2 \in \mathcal{U}$ are solutions to (5.5) with $\|\gamma - \varphi_i\|^* < \infty$ for $i = 1, 2$, then $\|\varphi_1 - \varphi_2\|^* < \infty$. So, it suffices to prove that $\varphi_1 = \varphi_2$ for every solutions $\varphi_1, \varphi_2 \in \mathcal{U}$ to (5.5) with

$$M := \|\varphi_1 - \varphi_2\|^* < \infty. \tag{5.26}$$

First consider the case where $p_i \in \mathbb{K}$ for $i = 1, \ldots, m$. The proof of uniqueness is by induction with respect to m. For $m = 1$ we have $p_1 = -a_0$, and therefore,

$$\mathcal{L}\varphi = F + p_1\varphi$$

for each $\varphi \in \mathcal{U}$ fulfilling equation (5.5).

Take solutions $\varphi_1, \varphi_2 \in \mathcal{U}$ to (5.5) that satisfy (5.26). Then we have

$$|p_1|^n \|\varphi_1 - \varphi_2\|^* = \|\mathcal{L}^n \varphi_1 - \mathcal{L}^n \varphi_2\|^* \leq \rho^n M, \qquad n \in \mathbb{N},$$

whence in the case $|p_1| > \rho$ we get $\varphi_1 = \varphi_2$. If $0 < |p_1| < \rho_0$, then (according to the assumptions) $\mathcal{L}$ is bijective,

$$\mathcal{L}^{-1}\varphi_i = p_1^{-1}\varphi_i - p_1^{-1}\mathcal{L}^{-1}F, \qquad i = 1, 2,$$

and consequently

$$|p_1|^{-n} \|\varphi_1 - \varphi_2\|^* = \|\mathcal{L}^{-n} \varphi_1 - \mathcal{L}^{-n} \varphi_2\|^* \leq \rho_0^{-n} M, \qquad n \in \mathbb{N},$$

which also implies that $\varphi_1 = \varphi_2$. If $p_1 = 0$ and $\mathcal{L}$ is bijective, then $\mathcal{L}\varphi_i = F$ for $i = 1, 2$, whence

$$\varphi_1 = \mathcal{L}^{-1}F = \varphi_2.$$

Now, fix $k \in \mathbb{N}$ and assume that the statement is true for $m = k$. We are to show that this is also the case for $m = k + 1$. So let $\varphi_1, \varphi_2 \in \mathcal{U}$ be solutions to (5.5) with $m = k + 1$ and (5.26) be valid. Write

$$\phi_i := \mathcal{L}\varphi_i - p_{k+1}\varphi_i, \qquad i = 1, 2.$$

Let $b_0, \ldots, b_{m-2} \in \mathbb{C}$ be as in Remark 19. Then

$$F = P_m(\mathcal{L})\varphi_i = \mathcal{L}^{k+1}\varphi_i + \sum_{j=0}^{k} a_j \mathcal{L}^j \varphi_i$$

$$= \mathcal{L}^k \phi_i + \sum_{j=0}^{k-1} b_j \mathcal{L}^j \phi_i, \qquad i = 1, 2.$$

So, ϕ_1 and ϕ_2 are solutions to equation (5.5), but with $m = k$ and a_i replaced by b_i for $i = 0, \ldots, k - 1$. Moreover, $p_1, \ldots, p_k$ are roots of the equation

$$z^k + \sum_{j=0}^{k-1} b_j z^j = 0,$$

which is the characteristic equation of the equation

$$\mathcal{L}^k \phi + \sum_{j=0}^{k-1} b_j \mathcal{L}^j \phi = F.$$

Note yet that, by the definition of ϕ_i,

$$\|\phi_1 - \phi_2\|^* \leq (\rho + |p_{k+1}|)M.$$

Hence, according to the inductive assumption, $\phi_1 = \phi_2$ and, analogously as in the case $m = 1$, finally we obtain that $\varphi_1 = \varphi_2$.

Now, to complete the proof, suppose that $p_i \notin \mathbb{K}$ for some $i \in \{1, \ldots, m\}$. Then, clearly, $\mathbb{K} = \mathbb{R}$. Let $\pi_1, \pi_2 : X^2 \to X$, $\widehat{\mathcal{U}}$, and $\widehat{\mathcal{L}} : \widehat{\mathcal{U}} \to \mathbb{C}(X)^S$ be given by (5.18)–(5.20). Then $\widehat{\mathcal{L}}(\widehat{\mathcal{U}}) \subset \widehat{\mathcal{U}}$, because $\mathcal{L}(\mathcal{U}) \subset \mathcal{U}$. Moreover (see Proposition 5), $\widehat{\mathcal{U}}$ is a linear subspace of $\mathbb{C}(X)$ over $\mathbb{C}$ and $\widehat{\mathcal{L}}$ is $\mathbb{C}$-linear.

Let $\varphi_1, \varphi_2 \in \mathcal{U}$ be solutions of (5.5) with

$$\|\varphi_1 - \varphi_2\|^* < \infty.$$

Define functions $\widehat{\varphi}_1, \widehat{\varphi}_2, F_0 : S \to \mathbb{C}(X)$ by

$$F_0(x) := (F(x), 0), \qquad \widehat{\varphi}_i(x) := (\varphi_i(x), 0), \qquad i = 1, 2, x \in S.$$

Note that $\widehat{\varphi}_1$ and $\widehat{\varphi}_2$ are solutions of the equation

$$\widehat{\mathcal{L}}^m \widehat{\phi} + \sum_{j=0}^{m-1} a_j \widehat{\mathcal{L}}^j \widehat{\phi} = F_0$$

and
$$\|\widehat{\varphi}_1 - \widehat{\varphi}_2\|_T^* < \infty.$$

Hence, by the first part of the proof, $\widehat{\varphi}_1 = \widehat{\varphi}_2$. Consequently $\varphi_1 = \varphi_2$.   □

Remark 21. In the case where $|p_i| < \|\mathcal{L}\|$ for some $i \in \{1, \ldots, m\}$, it follows from Theorem 95(c) (cf. [9, Theorem 3(c)]) that in the general situation the statement of Proposition 7 is not valid.

3. A general stability theorem

Now we show that, in some cases, the Hyers-Ulam stability of equation (5.5) can be derived from the analogous properties of the corresponding first-order equations, which we express in the form of the following hypothesis (for $i = 1, \ldots, m$):

$(\mathcal{H}_i)$ $p_i \in \mathbb{K}$ and $\rho_i : \mathbb{R}_+ \to \mathbb{R}_+$ is a function such that, for every $\varphi_s, \eta \in \mathcal{U}$ and $\delta \in \mathbb{R}_+$ with
$$\|\mathcal{L}\varphi_s - p_i\varphi_s - \eta\|^* \le \delta,$$

there is $\varphi \in \mathcal{U}$ such that
$$\mathcal{L}\varphi - p_i\varphi = \eta \tag{5.27}$$

and
$$\|\varphi_s - \varphi\|^* \le \rho_i(\delta).$$

For examples of operators satisfying $(\mathcal{H}_i)$ see the next sections.

The subsequent theorem and its proof have been patterned on [33, Theorem 2.1].

Theorem 89. *Let $\mathcal{L}(\mathcal{U}) \subset \mathcal{U}$ and $F \in X^S$. Assume that*
$(\mathcal{G})$ *$F \in \mathcal{U}$ or equation (5.5) has a solution $\widetilde{\varphi} \in \mathcal{U}_m$,*
$(\mathcal{H}_i)$ *holds for $i = 1, \ldots, m$, $\delta \in \mathbb{R}_+$, and $\varphi_s \in \mathcal{U}_m$ satisfies*
$$\|P_m(\mathcal{L})\varphi_s - F\|^* \le \delta. \tag{5.28}$$

Then there exists a solution $\varphi \in \mathcal{U}_m$ of equation (5.5) such that
$$\|\varphi_s - \varphi\|^* \le \rho_m \circ \ldots \circ \rho_1(\delta). \tag{5.29}$$

Moreover, if
$$|p_i| > \|\mathcal{L}\|, \qquad i = 1, \ldots, m,$$

then there is exactly one solution $\varphi \in \mathcal{U}$ of (5.5) with
$$\|\varphi_s - \varphi\|^* < \infty. \tag{5.30}$$

Proof. The idea of this proof is similar to some reasonings from the previous chapters, but for the convenience of readers we provide it.

Assume first that $F \in \mathcal{U}$. The proof is by induction with respect to m. To this end notice that the case $m = 1$ is a consequence of $(\mathcal{H}_1)$. So take $k \in \mathbb{N}$ and suppose that the theorem is true for $m = k$. We are to prove that it is true for $m = k + 1$.

To this end take $\varphi_s \in \mathcal{U}_m$ that satisfies (5.28) with $m = k + 1$, which by the Vieta formulas can be written in the form

$$\left\| \mathcal{L}^{k+1}\varphi_s + (-1)\left(\sum_{j=1}^{k+1} p_j\right)\mathcal{L}^k\varphi_s + \ldots + (-1)^{k+1}p_1 \ldots p_{k+1}\varphi_s - F \right\|^* \leq \delta.$$

Write

$$\psi_s := \mathcal{L}\varphi_s - p_{k+1}\varphi_s.$$

Then $\psi_s \in \mathcal{U}_k$. Since $\mathcal{L}$ is a linear operator, it is easily seen that

$$\mathcal{L}^p\psi_s = \mathcal{L}^{p+1}\varphi_s - p_{k+1}\mathcal{L}^p\varphi_s, \qquad p = 1, \ldots, k,$$

and consequently

$$\left\| \mathcal{L}^k\psi_s + (-1)\left(\sum_{j=1}^{k} p_j\right)\mathcal{L}^{k-1}\psi_s + \ldots + (-1)^k p_1 \ldots p_k\psi_s - F \right\|^*$$

$$= \left\| \mathcal{L}^{k+1}\varphi_s - p_{k+1}\mathcal{L}^k\varphi_s + (-1)\left(\sum_{j=1}^{k} p_j\right)\left(\mathcal{L}^k\varphi_s - p_{k+1}\mathcal{L}^{k-1}\varphi_s\right) \right.$$

$$\left. + \ldots + (-1)^k p_1 \ldots p_k\left(\mathcal{L}\varphi_s - p_{k+1}\varphi_s\right) - F \right\|^*$$

$$= \left\| \mathcal{L}^{k+1}\varphi_s + (-1)\left(\sum_{j=1}^{k+1} p_j\right)\mathcal{L}^k\varphi_s \right.$$

$$\left. + \ldots + (-1)^{k+1}p_1 \ldots p_{k+1}\varphi_s - F \right\|^* \leq \delta.$$

According to the inductive hypothesis there is $\psi \in \mathcal{U}_k$ with

$$\mathcal{L}^k\psi + (-1)\left(\sum_{j=1}^{k} p_j\right)\mathcal{L}^{k-1}\psi + \ldots + (-1)^k p_1 \ldots p_k\psi = F, \qquad (5.31)$$

$$\|\mathcal{L}\varphi_s - p_{k+1}\varphi_s - \psi\|^* = \|\psi_s - \psi\|^* \leq \rho_k \circ \ldots \circ \rho_1(\delta).$$

Hence, in view of $(\mathcal{H}_{k+1})$, there is $\varphi \in \mathcal{U}$ with

$$\mathcal{L}\varphi = p_{k+1}\varphi + \psi$$

and

$$\|\varphi_s - \varphi\|^* \leq \rho_{k+1}(\rho_k \circ \ldots \circ \rho_1(\delta)).$$

So, $\mathcal{L}\varphi \in \mathcal{U}$, which means that $\varphi \in \mathcal{U}_2$. Analogously, step by step, finally we get $\varphi \in \mathcal{U}_{k+1}$.

Clearly,

$$\psi = \mathcal{L}\varphi - p_{k+1}\varphi,$$

whence, by (5.31),

$$\begin{aligned}
F &= \mathcal{L}^{k+1}\varphi - p_{k+1}\mathcal{L}^k\varphi \\
&\quad + (-1)\left(\sum_{j=1}^{k} p_j\right)\left(\mathcal{L}^k\varphi - p_{k+1}\mathcal{L}^{k-1}\varphi\right) \\
&\quad + \ldots + (-1)^k p_1 \ldots p_k(\mathcal{L}\varphi - p_{k+1}\varphi) \\
&= \mathcal{L}^{k+1}\varphi + (-1)\left(\sum_{j=1}^{k+1} p_j\right)\mathcal{L}^k\varphi + \ldots + (-1)^{k+1} p_1 \ldots p_{k+1}\varphi \\
&= P_{k+1}(\mathcal{L})\varphi.
\end{aligned}$$

Now consider the case where equation (5.5) has a solution $\widetilde{\varphi} \in \mathcal{U}_m$. Let

$$\zeta_s := \varphi_s - \widetilde{\varphi}.$$

Then $\zeta_s \in \mathcal{U}_m$ and

$$\|P_m(\mathcal{L})\zeta_s - F_0\|^* = \|P_m(\mathcal{L})\zeta_s\|^* \leq \delta,$$

where $F_0 \in X^S$, $F_0(x) \equiv 0$. Clearly $F_0 \in \mathcal{U}$. Hence, by the first part of the proof, there exists a solution $\zeta \in \mathcal{U}_m$ of equation (5.5) (with $F = F_0$) such that

$$\|\zeta_s - \zeta\|^* \leq \rho_m \circ \ldots \circ \rho_1(\delta).$$

Now, it is easily seen that $\varphi := \zeta + \widetilde{\varphi}$ is a solution of (5.5) and (5.29) holds.

To complete the proof observe that the statement concerning the uniqueness of φ follows from Proposition 7. $\square$

Before the next theorem let us mention that we say that $\mathcal{U}$ is closed with respect to the uniform convergence (in the sequel shortly: uniformly closed) provided, for every sequence of functions $f_n \in \mathcal{U}$ $(n \in \mathbb{N})$ and $f \in X^S$ with

$$\lim_{n \to \infty} \|f - f_n\|^* = 0,$$

we have $f \in \mathcal{U}$.

Let us yet recall that a function $d : Y \times Y \to [0, +\infty]$ is called a generalized metric in a nonempty set Y if the following three properties are valid:

1) $d(x, y) = 0$ if and only if $x = y$;
2) $d(x, y) = d(y, x)$ for all $x, y \in Y$;
3) $d(x, z) \le d(x, y) + d(y, z)$ for all $x, y, z \in Y$.

The above concept differs from the usual one, of a metric space, by the fact that d may take also the infinite value. If d is a generalized metric on a nonempty set Y then we say that (Y, d) is a generalized metric space.

The next theorem is an analogue of the Banach's fixed point theorem for a generalized complete metric space; it is called the fixed point alternative of Diaz and Margolis [16, pp. 306-307].

Theorem 90. *Let (Y, d) be a generalized complete metric space and $\Lambda : Y \to Y$ be strictly contractive with the Lipschitz constant $L < 1$. If there exists $k \in \mathbb{N}$ such that $d(\Lambda^{k+1}x, \Lambda^k x) < \infty$ for some $x \in Y$, then the following properties are true.*

- *The sequence $(\Lambda^n x)_{n \in \mathbb{N}}$ converges to a fixed point $x^* \in Y$ of Λ.*
- *x^* is the unique fixed point of Λ in $Y^* = \{y \in Y : d(\Lambda^k x, y) < \infty\}$.*
- *If $y \in Y^*$, then $d(y, x^*) \le \dfrac{1}{1 - L} d(\Lambda y, y)$.*

Theorem 89 allows us to obtain the following two theorems. The first one is a small modification of [33, Theorem 2.3] (with a better estimation in (5.34) in the case of (β), because by Proposition 5, $\|\widehat{\mathcal{L}}\|_T \le \sqrt{2} \|\mathcal{L}\|$).

Theorem 91. *Assume that X is a Banach space, $\mathcal{U}$ is uniformly closed, $\mathcal{L}(\mathcal{U}) \subset \mathcal{U}$, $F \in X^S$, hypothesis $(\mathcal{G})$ of Theorem 89 is fulfilled, $\varphi_s \in \mathcal{U}$, and*

$$\delta := \|P_m(\mathcal{L})\varphi_s - F\|^* < \infty. \tag{5.32}$$

Let one of the following two conditions be valid:

(α)   *$p_i \in \mathbb{K}$ and $|p_i| > \|\mathcal{L}\|$ for $i = 1, \ldots, m$;*
(β)   *$|p_i| > \|\widehat{\mathcal{L}}\|_T$ for $i = 1, \ldots, m$.*
Then there is a unique solution $\varphi \in \mathcal{U}$ of the equation

$$P_m(\mathcal{L})\varphi = F \tag{5.33}$$

with $\|\varphi_s - \varphi\|^* < \infty$; *moreover,*

$$\|\varphi_s - \varphi\|^* \le \frac{\delta}{(|p_1| - \kappa) \dots (|p_m| - \kappa)}, \tag{5.34}$$

where

$$\kappa := \begin{cases} \|\mathcal{L}\|, & \text{if } (\alpha) \text{ holds;} \\ \|\widehat{\mathcal{L}}\|_T, & \text{if } (\beta) \text{ holds.} \end{cases}$$

Proof. First, consider the case of (α). In view of Theorem 89 it is enough to show that $(\mathcal{H}_i)$ holds for $i = 1, \dots, m$, with ρ_i given by (5.36).

Fix $i \in \{1, \dots, m\}$, $v, \psi \in \mathcal{U}$ and assume that

$$\delta_0 := \|\mathcal{L}\psi - p_i\psi - v\|^* < \infty.$$

Write

$$\mathcal{T}_i f := \frac{1}{p_i}(\mathcal{L}f - v), \qquad f \in \mathcal{U}. \tag{5.35}$$

Then

$$\|\mathcal{T}_i\psi - \psi\|^* \le \frac{\delta_0}{|p_i|},$$

and, for every $f, g \in \mathcal{U}$,

$$\|\mathcal{T}_i f - \mathcal{T}_i g\|^* = \left\| \frac{1}{p_i}\mathcal{L}f - \frac{1}{p_i}\mathcal{L}g \right\|^* \le \frac{\kappa}{|p_i|}\|f - g\|^*.$$

Define a generalized metric d in $\mathcal{U}$ by

$$d(f, g) = \|f - g\|^*, \qquad f, g \in \mathcal{U}.$$

Since $\mathcal{U}$ is uniformly closed, $(\mathcal{U}, d)$ is complete. Hence, by Theorem 90, the limit (with respect to d)

$$\phi := \lim_{n \to \infty} \mathcal{T}_i^n \psi$$

exists in $\mathcal{U}$ and ϕ is the unique fixed point of $\mathcal{T}_i$ with

$$\|\psi - \phi\|^* \le \frac{\delta_0}{|p_i|} \frac{1}{1 - \kappa/|p_i|} = \frac{\delta_0}{|p_i| - \kappa}.$$

Moreover,

$$\mathcal{L}\phi - p_i\phi = p_i(\mathcal{T}_i\phi - \phi) + v = v,$$

whence $(\mathcal{H}_i)$ is valid with

$$\rho_i(\delta_0) = \frac{\delta_0}{|p_i| - \kappa}, \qquad i = 1, \ldots, m, \ \delta_0 \in \mathbb{R}_+. \qquad (5.36)$$

This completes the proof when (α) is valid.

It remains to consider the case when (β) holds and $\mathbb{K} = \mathbb{R}$. To this end define π_1, π_2, $\widehat{\mathcal{U}}$ and $\widehat{\mathcal{L}}$ as in (5.18)–(5.20). Then (see Proposition 5) $\widehat{\mathcal{U}}$ is a linear subspace of $\mathfrak{C}(X)$ (over $\mathbb{C}$) and $\widehat{\mathcal{L}}$ is a linear operator (also over $\mathbb{C}$) such that $\widehat{\mathcal{L}}(\widehat{\mathcal{U}}) \subset \widehat{\mathcal{U}}$.

Let $\mu \in \mathfrak{C}(X)^S$ and $(\mu_n)_{n \in \mathbb{N}}$ be a sequence in $\widehat{\mathcal{U}}$ that is uniformly convergent to μ (in the Taylor norm). Then, by (5.17),

$$\max \{\|\pi_1 \circ \mu_n - \pi_1 \circ \mu\|^*, \|\pi_2 \circ \mu_n - \pi_2 \circ \mu\|^*\} \le \|\mu_n - \mu\|_T^*, \qquad n \in \mathbb{N},$$

which means that $(\pi_i \circ \mu_n)_{n \in \mathbb{N}}$ is uniformly convergent to $\pi_i \circ \mu$ for $i = 1, 2$. Consequently, $\pi_1 \circ \mu, \pi_2 \circ \mu \in \mathcal{U}$, whence $\mu \in \widehat{\mathcal{U}}$. Thus we have shown that $\widehat{\mathcal{U}}$ is uniformly closed with respect to the Taylor norm in $\mathfrak{C}(X)$.

Write

$$\chi := (\varphi_s, 0), \qquad F_0 := (F, 0).$$

Then

$$\left\|P_m(\widehat{\mathcal{L}})\chi - F_0\right\|_T^* = \left\|P_m(\mathcal{L})\varphi_s - F\right\|^* = \delta,$$

because

$$\pi_2 \circ \chi(x) \equiv 0.$$

Consequently, by the case of (α) (with $\kappa = \|\widehat{\mathcal{L}}\|_T$ and $\mathcal{L}$, φ_s, F, $\mathcal{U}$ replaced by $\widehat{\mathcal{L}}$, χ, F_0, $\widehat{\mathcal{U}}$, respectively), there is a solution $\mu \in \widehat{\mathcal{U}}_m$ of the equation

$$P_m(\widehat{\mathcal{L}})\mu = F_0 \qquad (5.37)$$

with

$$\|\chi - \mu\|_T^* \le \frac{\delta}{(|p_1| - \kappa) \ldots (|p_m| - \kappa)}. \qquad (5.38)$$

Observe that $\varphi := \pi_1 \circ \mu$ is a solution to (5.33). Moreover, by (5.38) and (5.17), we get (5.34).

To finish the proof it is enough to observe that the statement on uniqueness of φ results from Proposition 7. $\qquad \square$

For the next theorem let us recall (see (5.25)) that, if $\mathcal{L} : \mathcal{U} \to \mathcal{U}$ is bijective and $\mathcal{L}^{-1}$ is bounded, then

$$\frac{1}{\|\mathcal{L}^{-1}\|} \le \|\mathcal{L}\|.$$

Theorem 92. *Let X be a Banach space, $\mathcal{U}$ be uniformly closed, $\mathcal{L} : \mathcal{U} \to \mathcal{U}$ be bijective, $\|\mathcal{L}^{-1}\| < \infty$, $F \in \mathcal{U}$, and $\varphi_s \in \mathcal{U}_m$ satisfy (5.32). Assume that one of the following two conditions is valid:*

(α) $p_i \in \mathbb{K}$ *and* $|p_i| \notin [\, 1/\|\mathcal{L}^{-1}\|, \|\mathcal{L}\|\,]$ *for* $i = 1, \dots, m$;

(β) $|p_i| \notin \left[\, 1/\|\widehat{\mathcal{L}}^{-1}\|_T, \|\widehat{\mathcal{L}}\|_T \right]$ *for* $i = 1, \dots, m$.

Then there is a solution $\varphi \in \mathcal{U}_m$ of equation (5.33) with

$$\|\varphi_s - \varphi\|^* \le \frac{\delta}{\left|\kappa_1 - |p_1|\right| \dots \left|\kappa_m - |p_m|\right|}, \qquad (5.39)$$

where

$$\kappa_i := \begin{cases} \|\mathcal{L}\|, & \text{if } (\alpha) \text{ holds and } |p_i| > \|\mathcal{L}\|; \\ 1/\|\mathcal{L}^{-1}\|, & \text{if } (\alpha) \text{ holds and } |p_i| < 1/\|\mathcal{L}^{-1}\|; \\ \|\widehat{\mathcal{L}}\|_T, & \text{if } |p_i| > \|\widehat{\mathcal{L}}\|_T; \\ 1/\|\widehat{\mathcal{L}}^{-1}\|_T, & \text{if } |p_i| < 1/\|\widehat{\mathcal{L}}^{-1}\|_T. \end{cases}$$

Proof. First, consider the case of (α). We argue analogously as in the proof of Theorem 91. Namely, first we show that $(\mathcal{H}_i)$ holds for $i = 1, \dots, m$ with

$$\rho_i(\delta_0) = \frac{\delta_0}{\left|\kappa - |p_i|\right|}, \qquad \delta_0 \in \mathbb{R}_+,$$

and next use Theorem 89.

To this end fix $i \in \{1, \dots, m\}$. The case $|p_i| > \|\mathcal{L}\|$ follows from Theorem 91 (with $m = 1$). So, assume that $|p_i| < 1/\|\mathcal{L}^{-1}\|$. Fix $\varphi_s, \eta \in \mathcal{U}$ with

$$\delta_0 := \|\mathcal{L}\varphi_s - p_i\varphi_s - \eta\|^* < \infty. \qquad (5.40)$$

If $p_i = 0$, then for $\varphi := \mathcal{L}^{-1}\eta$ we have $\mathcal{L}\varphi - p_i\varphi = \mathcal{L}\varphi = \eta$ and

$$\|\varphi_s - \varphi\|^* = \|\varphi_s - \mathcal{L}^{-1}\eta\|^* \le \|\mathcal{L}^{-1}\| \, \|\mathcal{L}\varphi_s - \eta\|^*$$

$$\le \|\mathcal{L}^{-1}\|\delta_0 = \frac{\delta_0}{\kappa_i}.$$

If $p_i \ne 0$, we write $q_i = 1/p_i$, $\mathcal{T} = \mathcal{L}^{-1}$ and

$$\varphi_0 = \mathcal{L}\varphi_s. \qquad (5.41)$$

Then, by (5.40),

$$\|\mathcal{T}\varphi_0 - q_i\varphi_0 + q_i\eta\|^* = |q_i|\delta_0,$$

and $|q_i| > \|\mathcal{T}\| = \|\mathcal{L}^{-1}\|$. Consequently, by Theorem 91 (with $m = 1$), there is a function $\phi : S \to X$ such that

$$\mathcal{T}\phi - q_i\phi = -q_i\eta$$

and

$$\|\varphi_0 - \phi\|^* \leq \frac{|q_i|\delta_0}{|q_i| - \|\mathcal{L}^{-1}\|}.$$

Clearly, $\varphi := \mathcal{L}^{-1}\phi$ fulfills

$$\mathcal{L}\varphi - p_i\varphi = \eta,$$

$$\|\varphi_s - \varphi\|^* = \|\mathcal{L}^{-1}\varphi_0 - \mathcal{L}^{-1}\phi\|^* \leq \|\mathcal{L}^{-1}\| \|\varphi_0 - \phi\|^*$$

$$\leq \|\mathcal{L}^{-1}\|\frac{|q_i|\delta_0}{|q_i| - \|\mathcal{L}^{-1}\|} = \frac{\delta_0}{1/\|\mathcal{L}^{-1}\| - |p_i|}.$$

Thus we have proved that $(\mathcal{H}_i)$ is valid with

$$\rho_i(\delta_0) = \frac{\delta_0}{\kappa_i - |p_i|}, \qquad \delta_0 \in \mathbb{R}_+.$$

This completes the proof when (α) is valid.

The proof in the case of (β) runs similarly as in the analogous situation in the proof of Theorem 91. But for the convenience of readers we provide some details.

So assume that $\mathbb{K} = \mathbb{R}$ and define π_1, π_2, $\widehat{\mathcal{U}}$ and $\widehat{\mathcal{L}}$ as in (5.18)–(5.20). Then (see Proposition 5) $\widehat{\mathcal{U}}$ is a linear subspace of $\mathbb{C}(X)$ (over $\mathbb{C}$) and $\widehat{\mathcal{L}}$ is a linear operator (also over $\mathbb{C}$) such that $\widehat{\mathcal{L}}(\widehat{\mathcal{U}}) = \widehat{\mathcal{U}}$. Next, $\widehat{\mathcal{U}}$ is uniformly closed with respect to the Taylor norm in $\mathbb{C}(X)$ (see the proof of Theorem 91) and

$$\widehat{\mathcal{L}}^{-1}\mu(x) := (\mathcal{L}^{-1}(\pi_1 \circ \mu)(x), \mathcal{L}^{-1}(\pi_2 \circ \mu)(x)), \qquad \mu \in \widehat{\mathcal{U}}, x \in S.$$

If $\mu_s := (\varphi_s, 0)$ and $F_0 := (F, 0)$, then

$$\left\|P_m(\widehat{\mathcal{L}})\mu_s - F_0\right\|_T^* = \left\|P_m(\mathcal{L})\varphi_s - F\right\|^* = \delta < \infty,$$

and consequently, by the case of (α) (with $\mathcal{L}$, φ_s, F and $\mathcal{U}$ replaced by $\widehat{\mathcal{L}}$, μ_s, F_0 and $\widehat{\mathcal{U}}$, respectively), there is a solution $\mu \in \widehat{\mathcal{U}}$ of equation (5.37) with

$$\|\mu_s - \mu\|_T^* \leq \frac{\delta}{\left|\kappa_1 - |p_1|\right| \ldots \left|\kappa_m - |p_m|\right|}. \tag{5.42}$$

Clearly, $\varphi := \pi_1 \circ \mu$ is a solution to (5.33) and, by (5.17) and (5.42), we obtain that (5.39) holds. $\qquad\qquad\square$

The statement on uniqueness of φ does not need to hold in Theorem 92 in the general situation, as can be easily observed in Theorem 95 (a), (c).

4. Complementary results for the second-order equations

In this part we consider a particular case of (5.5) for $m = 2$ and $F(x) \equiv 0$, that is the equation

$$\mathcal{L}^2\varphi + a_1\mathcal{L}\varphi + a_0\varphi = 0, \tag{5.43}$$

with

$$a_0 \neq 0. \tag{5.44}$$

Let p_1 and p_2 denote, as before, the complex roots of the equation

$$z^2 + a_1 z + a_0 = 0,$$

which is the characteristic equation of (5.43). Then

$$a_1 = -p_1 - p_2, \qquad a_0 = p_1 p_2. \tag{5.45}$$

We assume that $a_1^2 - 4a_0 \neq 0$, that is

$$p_1 \neq p_2. \tag{5.46}$$

In what follows, given $g \in X^S$ and a sequence $(g_n)_{n \in \mathbb{N}}$ of functions from X^S, by the equality

$$g = \lim_{n \to \infty} g_n$$

we mean that

$$g(x) = \lim_{n \to \infty} g_n(x), \qquad x \in S.$$

Moreover, if $\mathcal{L}$ is injective, then we assume that $\mathcal{L}(\mathcal{U})$ is the domain of $\mathcal{L}^{-1}$, which means that $\mathcal{L}^{-1} : \mathcal{L}(\mathcal{U}) \to \mathcal{U}$ is bijective; clearly, in such a case we have $\|\mathcal{L}^{-1}\| \neq 0$.

The subsequent two theorems and their proofs have been patterned mainly on the results in [5], but they also generalize some earlier outcomes in [4, 20, 21].

Theorem 93. *Let X be a Banach space and $g \in \mathcal{U}_2$ satisfy the inequality*

$$\varepsilon := \left\|\mathcal{L}^2 g + a_1\mathcal{L}g + a_0 g\right\|^* < \infty. \tag{5.47}$$

Suppose that

$$p_1, p_2 \in \mathbb{K}, \tag{5.48}$$

(5.44) and (5.46) are valid, and one of the following three hypotheses holds:

(α) $\mathcal{L}(\mathcal{U}) \subset \mathcal{U}$, $\mathcal{U}$ is uniformly closed, and

$$L_1 := \|\mathcal{L}\| < |p_j|, \qquad j = 1, 2. \tag{5.49}$$

(β) $\mathcal{L}$ is injective, $\mathcal{L}(\mathcal{U})$ is uniformly closed, $\mathcal{U} \subset \mathcal{L}(\mathcal{U})$, and

$$|p_j| < \|\mathcal{L}^{-1}\|^{-1} =: L_2, \qquad j = 1, 2. \qquad (5.50)$$

(γ) $\mathcal{L}$ is injective, $L_1 < |p_1|, |p_2| < L_2$, $\mathcal{L}(\mathcal{U}) = \mathcal{U}$ and $\mathcal{U}$ is uniformly closed.

Then there exists a unique solution $h \in \mathcal{U}_2$ of equation (5.43) with

$$\|g - h\|^* < \infty; \qquad (5.51)$$

moreover, h is given by (5.74) and

$$\|g - h\|^* \le \frac{\varepsilon}{|p_2 - p_1|} \left(\frac{1}{|L_1' - |p_1||} + \frac{1}{|L_2' - |p_2||} \right), \qquad (5.52)$$

where

$$L_i' := \begin{cases} L_1, & if\ (\alpha)\ holds; \\ L_2, & if\ (\beta)\ holds; \\ L_i, & if\ (\gamma)\ holds, \end{cases} \qquad i \in \{1, 2\}. \qquad (5.53)$$

For the proof of Theorem 93 we need the following lemmas.

Lemma 7. *Assume that (5.44), (5.46), and (5.48) are valid and one of hypotheses (α)–(γ) is fulfilled. Let $f_1, f_2 \in \mathcal{U}_2$ be solutions of equation (5.43) and $\|f_1 - f_2\|^* < \infty$. Then $f_1 = f_2$.*

Proof. The case of (α) can be actually deduced from Proposition 7. However, since in that situation the reasonings are analogous as for (β) and (γ), we do not skip that case in this proof.

Note that $a_1 = -p_1 - p_2$ and $a_0 = p_1 p_2$. So, by (5.44), $p_i \ne 0$ for $i = 1, 2$. Let $h_i := \mathcal{L}f_i - p_2 f_i$ for $i \in \{1, 2\}$. Then

$$\mathcal{L}h_i = \mathcal{L}^2 f_i - p_2 \mathcal{L}f_i = -a_1 \mathcal{L}f_i - a_0 f_i - p_2 \mathcal{L}f_i$$
$$= p_1(\mathcal{L}f_i - p_2 f_i) = p_1 h_i$$

for $i \in \{1, 2\}$. Hence, if (α) or (γ) holds,

$$\|h_1 - h_2\|^* = |p_1|^{-k} \|\mathcal{L}^k h_1 - \mathcal{L}^k h_2\|^* \le |p_1|^{-k} L_1^k \|h_1 - h_2\|^*, \qquad k \in \mathbb{N};$$

if (β) holds,

$$\|h_1 - h_2\|^* = |p_1|^k \|\mathcal{L}^{-k} h_1 - \mathcal{L}^{-k} h_2\|^* \le |p_1|^k L_2^{-k} \|h_1 - h_2\|^*, \qquad k \in \mathbb{N}.$$

Consequently, with $k \to \infty$ we get $h_1 = h_2$. Next,

$$\mathcal{L}f_1 - p_2 f_1 - (\mathcal{L}f_2 - p_2 f_2) = h_1 - h_2 = 0,$$

whence

$$\mathcal{L}f_1 - \mathcal{L}f_2 = p_2(f_1 - f_2),$$

and consequently, for each $k \in \mathbb{N}$, in the case of (α) we have

$$\|f_1 - f_2\|^* = |p_2|^{-k}\|\mathcal{L}^k f_1 - \mathcal{L}^k f_2\|^* \le |p_2|^{-k}L_1^k\|f_1 - f_2\|^*,$$

and, in the case of (β) or (γ),

$$\|f_1 - f_2\|^* = |p_2|^k\|\mathcal{L}^{-k}f_1 - \mathcal{L}^{-k}f_2\|^* \le |p_2|^k L_2^{-k}\|f_1 - f_2\|^*.$$

Again with $k \to \infty$ we obtain $f_1 = f_2$. □

Lemma 8. *Assume that (5.44), (5.46), and (5.48) are valid, one of hypotheses (α), (γ) is fulfilled, and $L_1 < |p_i|$. Let $g \in \mathcal{U}_2$, (5.47) be valid, $i \in \{1, 2\}$, and*

$$\mathcal{J}_k^i := p_i^{-k}[\mathcal{L}^k g + (a_1 a_0^{-1} + p_i^{-1})\mathcal{L}^{k+1}g], \qquad k \in \mathbb{N}_0.$$

Then the limit

$$F_i(x) = \lim_{n \to \infty} \mathcal{J}_n^i(x) \tag{5.54}$$

exists for each $x \in S$.

Moreover, if $F_i : S \to X$ is defined by (5.54) for each $x \in S$, then

$$F_i \in \mathcal{U}_2, \tag{5.55}$$

$$\left\|g + (a_1 a_0^{-1} + p_i^{-1})\mathcal{L}g - F_i\right\|^* \le \frac{|a_0|^{-1}|p_i|\varepsilon}{|p_i| - L_1}, \tag{5.56}$$

$$\mathcal{L}^2 F_i + a_1 \mathcal{L}F_i + a_0 F_i = 0. \tag{5.57}$$

Proof. According to the assumptions, $\mathcal{U}$ is uniformly closed,

$$L_1 |p_i|^{-1} < 1, \qquad \mathcal{L}(\mathcal{U}) \subset \mathcal{U}. \tag{5.58}$$

Clearly, $\mathcal{J}_k^i \in \mathcal{U}$ for all $k \in \mathbb{N}_0$ and, by (5.45),

$$p_i^{-1}(a_1 a_0^{-1} + p_i^{-1}) = -a_0^{-1}, \tag{5.59}$$

whence, for each $k \in \mathbb{N}_0$,

$$
\begin{aligned}
\mathcal{J}_k^i - \mathcal{J}_{k+1}^i &= p_i^{-k}[\mathcal{L}^k g + (a_1 a_0^{-1} + p_i^{-1})\mathcal{L}^{k+1}g] \\
&\quad - p_i^{-k-1}[\mathcal{L}^{k+1}g + (a_1 a_0^{-1} + p_i^{-1})\mathcal{L}^{k+2}g] \\
&= p_i^{-k}\mathcal{L}^k(g + a_1 a_0^{-1}\mathcal{L}g + a_0^{-1}\mathcal{L}^2 g),
\end{aligned}
$$

and consequently, by (5.47),

$$\|\mathcal{J}_k^i - \mathcal{J}_{k+1}^i\|^* \leq |p_i|^{-k}L_1^k\|g + a_1 a_0^{-1}\mathcal{L}g + a_0^{-1}\mathcal{L}^2 g\|^*$$
$$\leq |p_i|^{-k}L_1^k \varepsilon |a_0|^{-1},$$

which implies that

$$\|\mathcal{J}_k^i - \mathcal{J}_{k+n}^i\|^* \leq \varepsilon |a_0|^{-1} \sum_{j=k}^{k+n-1} |p_i|^{-j}L_1^j, \qquad n \in \mathbb{N}. \tag{5.60}$$

Consequently, by (5.58), the sequence $(\mathcal{J}_n^i(x))_{n\in\mathbb{N}}$ is Cauchy for each $x \in S$. Therefore limit (5.54) exists for each $x \in S$. Next, by (5.60), $\mathcal{J}_n^i$ tends uniformly to F_i, whence we have (5.55) (see (5.58)).

Note that

$$\|\mathcal{L}^j F_i - \mathcal{L}^j \mathcal{J}_k^i\|^* \leq L_1^j \|F_i - \mathcal{J}_k^i\|^*, \qquad k \in \mathbb{N}, j = 1, 2,$$

and consequently, by the definition of F_i,

$$\mathcal{L}^j F_i = \lim_{k\to\infty} \mathcal{L}^j \mathcal{J}_k^i, \qquad j \in \{1,2\}, k \in \mathbb{N}.$$

Further,

$$\mathcal{L}^n \mathcal{J}_k^i = p_i^n \mathcal{J}_{n+k}^i, \qquad n, k \in \mathbb{N}_0. \tag{5.61}$$

Hence,

$$\begin{aligned} \mathcal{L}^2 F_i + a_1 \mathcal{L}F_i &= \lim_{n\to\infty} \mathcal{L}^2 \mathcal{J}_n^i + a_1 \lim_{n\to\infty} \mathcal{L}\mathcal{J}_n^i \\ &= p_i^2 \lim_{n\to\infty} \mathcal{J}_{n+2}^i + a_1 p_i \lim_{n\to\infty} \mathcal{J}_{n+1}^i \\ &= p_i^2 F_i + a_1 p_i F_i = -a_0 F_i, \end{aligned} \tag{5.62}$$

which gives (5.57). Finally, by (5.60) with $k = 0$ and $n \to \infty$, we obtain (5.56).   □

Lemma 9. *Assume that (5.44), (5.46) and (5.48) hold, one of hypotheses (β), (γ) is fulfilled, and $L_2 > |p_i|$. Let $g \in \mathcal{U}_2$, (5.47) be valid, $i \in \{1,2\}$, and*

$$\mathcal{J}_k^i := p_i^k [\mathcal{L}^{-k}g + (a_1 a_0^{-1} + p_i^{-1})\mathcal{L}^{-k+1}g], \qquad k \in \mathbb{N}_0.$$

Then the limit

$$F_i(x) = \lim_{n\to\infty} \mathcal{J}_n^i(x) \tag{5.63}$$

exists for each $x \in S$.

Moreover, if $F_i : S \to X$ is defined by (5.63) for each $x \in S$, then

$$F_i \in \mathcal{U}_2, \tag{5.64}$$

$$\left\| g + (a_1 a_0^{-1} + p_i^{-1})\mathcal{L}g - F_i \right\|^* \le \frac{|a_0|^{-1}|p_i|\varepsilon}{L_2 - |p_i|}, \tag{5.65}$$

$$\mathcal{L}^2 F_i + a_1 \mathcal{L} F_i + a_0 F_i = 0. \tag{5.66}$$

Proof. According to the assumptions, $\mathcal{L}$ is injective, $\mathcal{L}(\mathcal{U})$ is uniformly closed,

$$|p_i| L_2^{-1} < 1, \tag{5.67}$$

and

$$\mathcal{L}^{-k}(\mathcal{U}) \subset \mathcal{U} \subset \mathcal{L}(\mathcal{U}), \qquad k \in \mathbb{N}. \tag{5.68}$$

Next,

$$\mathcal{J}_k^i \in \mathcal{U}, \qquad k \in \mathbb{N}, \tag{5.69}$$

(because $g \in \mathcal{U}_2 \subset \mathcal{U}$) and, by (5.59),

$$\begin{aligned}
\mathcal{J}_k^i - \mathcal{J}_{k-1}^i &= p_i^k [\mathcal{L}^{-k} g + (a_1 a_0^{-1} + p_i^{-1})\mathcal{L}^{-k+1} g] \\
&\quad - p_i^{k-1}[\mathcal{L}^{-k+1} g + (a_1 a_0^{-1} + p_i^{-1})\mathcal{L}^{-k+2} g] \\
&= p_i^k \mathcal{L}^{-k}(g + a_1 a_0^{-1} \mathcal{L}g + a_0^{-1}\mathcal{L}^2 g), \qquad k \in \mathbb{N},
\end{aligned}$$

whence

$$\begin{aligned}
\|\mathcal{J}_k^i - \mathcal{J}_{k-1}^i\|^* &\le |p_i|^k L_2^{-k}\|g + a_1 a_0^{-1}\mathcal{L}g + a_0^{-1}\mathcal{L}^2 g\|^* \\
&\le |a_0|^{-1}|p_i|^k L_2^{-k}\varepsilon, \qquad k \in \mathbb{N}.
\end{aligned}$$

Consequently,

$$\|\mathcal{J}_k^i - \mathcal{J}_{k+n}^i\|^* \le |a_0|^{-1}\varepsilon \sum_{j=k+1}^{k+n} |p_i|^j L_2^{-j}, \qquad k, n \in \mathbb{N}_0, \, n > 0. \tag{5.70}$$

Thus (5.67) yields that the sequence $(\mathcal{J}_n^i(x))_{n \in \mathbb{N}}$ is Cauchy for each $x \in S$ and therefore there exists

$$F_i(x) = \lim_{n \to \infty} \mathcal{J}_n^i(x), \qquad x \in S.$$

Since, by (5.70), $\mathcal{J}_n^i$ tends uniformly to F_i, (5.68) and (5.69) hold, and $\mathcal{L}(\mathcal{U})$ is uniformly closed, we have $F_i \in \mathcal{L}(\mathcal{U})$ and from (5.68) we deduce further that

$$F_i \in \mathcal{L}^n(\mathcal{U}), \qquad n \in \mathbb{N}. \tag{5.71}$$

Next, letting $n \to \infty$, with $k = 0$, we deduce that

$$\left\| g + (a_1 a_0^{-1} + p_i^{-1})\mathcal{L}g - F_i \right\|^* \leq \frac{|a_0|^{-1}\varepsilon L_2^{-1}|p_i|}{1 - L_2^{-1}|p_i|}$$
$$= \frac{|a_0|^{-1}|p_i|\varepsilon}{L_2 - |p_i|},$$

which is (5.65).

Next, note that

$$\mathcal{L}^{-n}\mathcal{J}_k^i = p_i^{-n}\mathcal{J}_{k+n}^i, \qquad k, n \in \mathbb{N}_0, \tag{5.72}$$

and (5.50) implies that

$$\mathcal{L}^{-k}F_i = \lim_{n\to\infty} \mathcal{L}^{-k}\mathcal{J}_n^i, \qquad k \in \mathbb{N}.$$

So, (5.72) yields that

$$a_1 \mathcal{L}^{-1}F_i = a_1 \lim_{n\to\infty} \mathcal{L}^{-1}\mathcal{J}_n^i$$
$$= a_1 p_i^{-1} \lim_{n\to\infty} \mathcal{J}_{n+1}^i = a_1 p_i^{-1}F_i$$

and therefore

$$a_1 \mathcal{L}^{-1}F_i + F_i = a_1 p_i^{-1}F_i + F_i$$
$$= -a_0 p_i^{-2}F_i = -a_0 \lim_{n\to\infty} p_i^{-2}\mathcal{J}_n^i$$
$$= -a_0 \lim_{n\to\infty} \mathcal{L}^{-2}\mathcal{J}_{n-2}^i = -a_0\mathcal{L}^{-2}F_i.$$

From (5.71) we deduce that

$$F_i = -a_1\mathcal{L}^{-1}F_i - a_0\mathcal{L}^{-2}F_i \in \mathcal{U}.$$

Repeating that reasoning we obtain that

$$F_i \in \mathcal{L}^{-1}(\mathcal{U}),$$

whence (5.64) holds and

$$\mathcal{L}^2 F_i + a_1\mathcal{L}F_i = \mathcal{L}^2(a_1\mathcal{L}^{-1}F_i + F_i)$$
$$= \mathcal{L}^2(-a_0\mathcal{L}^{-2}F_i) = -a_0 F_i,$$

which is (5.66). $\square$

Now, we are in a position to prove Theorem 93.

Proof of Theorem 93. In view of the assumptions, one of hypotheses (α)–(γ) is fulfilled and therefore, for each $i \in \{1, 2\}$, either $|p_i| > L_1$ or $|p_i| < L_2$. In either case, according to Lemmas 8 and 9, the suitable defined function F_i satisfies equation (5.43) and one of inequalities (5.56), (5.65), respectively, holds; moreover, by (5.55) and (5.64),

$$F_i \in \mathcal{U}_2, \qquad i = 1, 2. \tag{5.73}$$

Let $h : S \rightarrow X$ be given by

$$h := \frac{p_2}{p_2 - p_1} F_1 - \frac{p_1}{p_2 - p_1} F_2. \tag{5.74}$$

It is easy to see that (5.73) implies

$$h \in \mathcal{U}_2$$

and, by (5.57) and (5.66), respectively,

$$a_1 \mathcal{L} h + \mathcal{L}^2 h = \frac{p_2}{p_2 - p_1} [a_1 \mathcal{L} F_1 + \mathcal{L}^2 F_1] - \frac{p_1}{p_2 - p_1} [a_1 \mathcal{L} F_2 + \mathcal{L}^2 F_2]$$

$$= \frac{p_2}{p_2 - p_1}(-a_0 F_1) - \frac{p_1}{p_2 - p_1}(-a_0 F_2) = -a_0 h.$$

Next, it is easy to check that

$$g - h = \frac{1}{p_2 - p_1}((p_2 - p_1)g - p_2 F_1 + p_1 F_2)$$

and

$$p_2(a_1 a_0^{-1} + p_1^{-1}) - p_1(a_1 a_0^{-1} + p_2^{-1}) = 0,$$

whence

$$\|g - h\|^* = \frac{1}{|p_2 - p_1|}\|(p_2 - p_1)g - p_2 F_1 + p_1 F_2$$

$$+ [p_2(a_1 a_0^{-1} + p_1^{-1}) - p_1(a_1 a_0^{-1} + p_2^{-1})]\mathcal{L} g\|^*$$

$$\leq \frac{|p_2|}{|p_2 - p_1|}\|g + (a_1 a_0^{-1} + p_1^{-1})\mathcal{L} g - F_1\|^*$$

$$+ \frac{|p_1|}{|p_2 - p_1|}\|g + (a_1 a_0^{-1} + p_2^{-1})\mathcal{L} g - F_2\|^*.$$

So, if $L_1 < |p_i|$ for $i = 1, 2$, then (5.56) yields

$$\|g - h\|^* \leq \frac{|p_2|}{|p_2 - p_1|} \frac{|a_0|^{-1}|p_1|\varepsilon}{|p_1| - L_1} + \frac{|p_1|}{|p_2 - p_1|} \frac{|a_0|^{-1}|p_2|\varepsilon}{|p_2| - L_1} \tag{5.75}$$

$$= \frac{\varepsilon}{|p_2 - p_1|}\left(\frac{1}{|p_1| - L_1} + \frac{1}{|p_2| - L_1}\right);$$

while, with $L_2 > |p_i|$ for $i = 1, 2$, from (5.65) we get

$$\|g - h\|^* \leq \frac{\varepsilon}{|p_2 - p_1|} \left(\frac{1}{L_2 - |p_1|} + \frac{1}{L_2 - |p_2|} \right). \tag{5.76}$$

Further, in the case when $L_1 < |p_1|$ and $L_2 > |p_2|$, conditions (5.56) and (5.65), respectively, imply that

$$\|g - h\|^* \leq \frac{\varepsilon}{|p_2 - p_1|} \left(\frac{1}{|p_1| - L_1} + \frac{1}{L_2 - |p_2|} \right). \tag{5.77}$$

Thus we have proved (5.52).

We are yet to show the uniqueness of h. So, take a solution $h_0 \in \mathcal{U}_2$ of (5.43) with $\|g - h_0\|^* < \infty$. Clearly,

$$\|h - h_0\|^* \leq \|h - g\|^* + \|g - h_0\|^* < \infty,$$

whence Lemma 7 implies $h = h_0$. This ends the proof of Theorem 93.

Below, we provide four simple and natural examples of linear operators $\mathcal{L}$ that satisfy the assumptions of Theorem 93, with suitable p_1, p_2 (cf. [5]).

Example 10. Let $\mathcal{U} = X^S$, $n \in \mathbb{N}$, and

$$\mathcal{L}f = \sum_{i=1}^{n} \Psi_i \circ f \circ \xi_i,$$

with some $\xi_1, \ldots, \xi_n : S \to S$ and bounded linear $\Psi_1, \ldots, \Psi_n : X \to X$. Then

$$\|\mathcal{L}f(x) - \mathcal{L}h(x)\| \leq \sum_{i=1}^{n} \lambda_i \|f(\xi_i(x)) - h(\xi_i(x))\|, \qquad f, h \in X^S, \ x \in S,$$

with

$$\lambda_i = \|\Psi_i\|, \qquad i = 1, \ldots, m,$$

and consequently

$$\|\mathcal{L}\| \leq \sum_{i=1}^{n} \lambda_i.$$

Example 11. Let $\mathcal{U} = X^S$, $\Psi : X \to X$ and $\xi : S \to S$ be bijections, Ψ be linear, Ψ^{-1} be bounded, and

$$\mathcal{L}f = \Psi \circ f \circ \xi, \qquad f \in X^S.$$

Then

$$\mathcal{L}^{-1} f = \Psi^{-1} \circ f \circ \xi^{-1}, \qquad f \in X^S,$$

whence

$$\left\| \mathcal{L}^{-1} f(x) \right\| \leq \left\| \Psi^{-1} \right\| \left\| f(\xi^{-1}(x)) \right\|, \qquad f \in X^S, \ x \in S.$$

Hence

$$0 < \left\| \mathcal{L}^{-1} \right\| \leq \left\| \Psi^{-1} \right\|.$$

Example 12. Let $a, b \in \mathbb{R}$, $a < b$, $S = [a, b]$, $\mathcal{U}$ be the family of all continuous functions mapping the interval $[a, b]$ into $\mathbb{R}$, $n \in \mathbb{N}$, $A_1, \ldots, A_n \in \mathbb{R}$, $\xi_1, \ldots, \xi_n : S \to S$ be continuous and

$$\mathcal{L} f(x) = \sum_{i=1}^n \int_a^x A_i f \circ \xi_i(t) dt$$

for $f \in \mathcal{U}$, $x \in S$. Then

$$\|\mathcal{L}\| \leq (b - a) \sum_{i=1}^n |A_i|.$$

Example 13. Let $a, b \in \mathbb{R}$, $a < b$, $\mathcal{U}$ be the family of all continuously differentiable functions $f : [a, b] \to \mathbb{R}$ with $f(a) = 0$, and $\mathcal{L} = d/dt$. Then

$$\|\mathcal{L}^{-1}\| := b - a.$$

In the next theorem, we consider the case when $\mathbb{K} = \mathbb{R}$ and (5.48) is not fulfilled, which complements Theorem 93.

To this end let us remind that $\mathfrak{C}(X)$ denotes the complexification of X and $\pi_1, \pi_2 : X^2 \to X$ are defined by (5.18). Let $\widehat{\mathcal{U}}$ and $\widehat{\mathcal{L}} : \widehat{\mathcal{U}} \to \mathfrak{C}(X)$ be given by (5.19) and (5.20). Then, by Proposition 5, $\widehat{\mathcal{U}}$ is a linear subspace of $\mathfrak{C}(X)$ (over $\mathbb{C}$), $\widehat{\mathcal{L}}$ is $\mathbb{C}$-linear, and

$$\|\widehat{\mathcal{L}}\|_T \leq \sqrt{2} \, \|\mathcal{L}\|.$$

Let us mention yet that the subsequent theorem is more general than [5, Theorem 4.1] and with a better estimate in (5.81) (analogous as in Theorem 93).

Theorem 94. *Let X be a Banach space, $g \in \mathcal{U}_2$, and (5.47) hold. Suppose that (5.44) and (5.46) are valid, and one of the following three hypotheses is fulfilled.*
(α) $\mathcal{L}(\mathcal{U}) \subset \mathcal{U}$, $\mathcal{U}$ is uniformly closed, and

$$\widehat{L}_1 := \|\widehat{\mathcal{L}}\|_T < |p_j|, \qquad j = 1, 2. \tag{5.78}$$

(β) $\mathcal{L}$ is injective, $\mathcal{L}(\mathcal{U})$ is uniformly closed, $\mathcal{U} \subset \mathcal{L}(\mathcal{U})$, and

$$|p_j| < \left\|\widehat{\mathcal{L}^{-1}}\right\|_T^{-1} =: \widehat{L}_2, \qquad j = 1, 2. \tag{5.79}$$

(γ) $\mathcal{L}$ is injective, $\widehat{L}_1 < |p_1|$, $\widehat{L}_2 > |p_2|$, $\mathcal{L}(\mathcal{U}) = \mathcal{U}$ and $\mathcal{U}$ is uniformly closed.
Then there exists a unique solution $H \in \mathcal{U}_2$ of equation (5.43) with

$$\|g - H\|^* < \infty; \tag{5.80}$$

moreover,

$$\|g - H\|^* \leq \frac{\varepsilon}{|p_2 - p_1|}\left(\frac{1}{\left|\widehat{L}_1' - |p_1|\right|} + \frac{1}{\left|\widehat{L}_2' - |p_2|\right|}\right), \tag{5.81}$$

where

$$\widehat{L}_i' := \begin{cases} \widehat{L}_1, & \text{if } (\alpha) \text{ holds;} \\ \widehat{L}_2, & \text{if } (\beta) \text{ holds;} \\ \widehat{L}_i, & \text{if } (\gamma) \text{ holds,} \end{cases} \qquad i \in \{1, 2\}. \tag{5.82}$$

Proof. Define $\chi : S \to X^2$ by

$$\chi(x) := (g(x), 0), \qquad x \in S.$$

Since $\mathcal{L}(\mathcal{U}) \subset \mathcal{U}$ and $\mathcal{U}$ is a real linear subspace of X^S, it is easily seen that

$$\widehat{\mathcal{L}}(\widehat{\mathcal{U}}) \subset \widehat{\mathcal{U}} \tag{5.83}$$

and $\chi, \widehat{\mathcal{L}}\chi, \widehat{\mathcal{L}}^2\chi \in \widehat{\mathcal{U}}$.

   Next, analogously as in the proof of Theorem 91, we show that $\widehat{\mathcal{U}}$ (or $\widehat{\mathcal{L}}(\widehat{\mathcal{U}})$ in the case of (β)) is uniformly closed (with regard to the Taylor norm).

   Note yet that, according to (5.17), we have

$$\left\|\widehat{\mathcal{L}}^2\chi + a_1\widehat{\mathcal{L}}\chi + a_0\chi\right\|_T^* = \left\|(\mathcal{L}^2 g + a_1\mathcal{L}g + a_0 g, 0)\right\|_T^*$$
$$= \left\|\mathcal{L}^2 g + a_1\mathcal{L}g + a_0 g\right\|^* = \varepsilon,$$

because

$$\pi_2 \circ \chi(x) = 0, \qquad x \in S.$$

   So, the assumptions of Theorem 93 are satisfied (with g, $\mathcal{L}$, and $\mathcal{U}$ replaced by χ, $\widehat{\mathcal{L}}$, and $\widehat{\mathcal{U}}$, respectively), and consequently, there is a $\eta \in \widehat{\mathcal{U}}$, with $\widehat{\mathcal{L}}\eta \in \widehat{\mathcal{U}}$, which is a solution of the equation

$$\widehat{\mathcal{L}}^2\eta + a_1\widehat{\mathcal{L}}\eta + a_0\eta = 0$$

and fulfills the inequality

$$\|\chi - \eta\|_T^* \le \frac{\varepsilon}{|p_2 - p_1|}\left(\frac{1}{\left|\widehat{L_1'} - |p_1|\right|} + \frac{1}{\left|\widehat{L_2'} - |p_2|\right|}\right).$$

Clearly, $H := \pi_1 \circ \eta$ is a solution of (5.43) and, by (5.17), (5.81) holds.

It remains to show the uniqueness of H. So, let $H_0 \in \mathcal{U}_2$ be a solution of equation (5.43) such that $\|g - H_0\|^* < \infty$. Write

$$\mu(x) := (H_0(x), \pi_2(\eta(x))), \qquad x \in S.$$

It is easily seen that $\mu \in \widehat{\mathcal{U}}$ and

$$\widehat{\mathcal{L}}^2\mu + a_1\widehat{\mathcal{L}}\mu + a_0\mu = 0,$$

because $\pi_2 \circ \eta$ is a solution to (5.43), too. Moreover,

$$\|\eta - \mu\|_T^* = \|H - H_0\|^* \le \|H - g\|^* + \|g - H_0\|^* < \infty.$$

Hence, by Lemma 7 (with $\mathcal{L}$ and $\mathcal{U}$ replaced by $\widehat{\mathcal{L}}$ and $\widehat{\mathcal{U}}$, respectively), $\eta = \mu$, which yields

$$H_0 = \pi_1 \circ \mu = \pi_1 \circ \eta = H.$$

$\square$

5. Linear difference equation with constant coefficients

Now, consider stability of a simple particular case of equation (5.5), that is the linear difference equation

$$\varphi(n + m) + a_{m-1}\varphi(n + m - 1) + \cdots + a_1\varphi(n + 1) + a_0\varphi(n) = F(n), \qquad (5.84)$$

with $S \in \{\mathbb{Z}, \mathbb{N}\}$, the given function $F : S \to X$ and the unknown function $\varphi : S \to X$. Clearly, (5.5) becomes (5.84) with $\mathcal{L}\varphi(n) = \varphi(n + 1)$ for $n \in S$. If we write $\varphi_n = \varphi(n)$ and $F_n := F(n)$ for $n \in S$, then (5.84) takes the form

$$\varphi_{n+m} + a_{m-1}\varphi_{n+m-1} + \cdots + a_0\varphi_n = F_n, \qquad n \in S, \qquad (5.85)$$

which we will use in the sequel.

The Ulam stability of it has already been investigated in several papers (see, e.g., [6, 7, 9, 10, 28, 29]; cf. also [11, 12]). In this part we present some results that can be derived from Theorems 89, 91, 93, and 94. In what follows,

$$\mathbb{S} := \{a \in \mathbb{C} : |a| = 1\}.$$

We start with the following (see [9]):

Theorem 95. *Let $T \in \{\mathbb{N}_0, \mathbb{Z}\}$, $\delta > 0$ and $(y_n)_{n \in T}$, $(b_n)_{n \in T}$ be two sequences in X such that*

$$\|y_{n+m} + a_{m-1}y_{n+m-1} + \ldots + a_0 y_n - b_n\| \leq \delta, \qquad n \in T. \tag{5.86}$$

Assume that one of the following two hypotheses is valid.

(α) $T = \mathbb{N}_0$ and $|p_i| < 1$ for $i = 1, \ldots, m$.

(β) X is a Banach space and $|p_i| \neq 1$ for $i = 1, \ldots, m$.

Then there exists a sequence $(x_n)_{n \in T}$ in X such that

$$x_{n+m} + a_{m-1}x_{n+m-1} + \ldots + a_0 x_n = b_n, \qquad n \in T, \tag{5.87}$$

and

$$\|y_n - x_n\| \leq \frac{\delta}{\left|1 - |p_1|\right| \ldots \left|1 - |p_m|\right|}, \qquad n \in T. \tag{5.88}$$

Moreover, the following three statements are valid.

(a) $(x_n)_{n \in T}$ is unique if and only if one of the following two conditions is valid

$$|p_i| > 1, \qquad i = 1, \ldots, m, \tag{5.89}$$

$$T = \mathbb{Z}. \tag{5.90}$$

(b) If either (5.89) or (5.90) holds, then $(x_n)_{n \in T}$ is the unique sequence in X, such that (5.87) is valid and

$$\sup_{n \in T} \|x_n - y_n\| < \infty.$$

(c) If either (5.89) does not hold or $T = \mathbb{N}_0$, then the cardinality of the set of all sequences $(x_n)_{n \in T}$ in X, satisfying (5.87) and (5.88), equals the cardinality of X.

Proof. We start with a proof of existence of $(x_n)_{n \in T}$.

Note that (5.87) is a particular case of (5.5) with $S = T$ and

$$\mathcal{L}f(n) := f(n + 1), \qquad n \in T.$$

This means that, in the case $T = \mathbb{Z}$, $\mathcal{L}$ is bijective and

$$\mathcal{L}^{-1}f(n) := f(n - 1), \qquad n \in T.$$

Further, it is easily seen that $\|\mathcal{L}\| = 1 = \|\widehat{\mathcal{L}}\|_T$ and $\|\mathcal{L}^{-1}\| = 1 = \|\widehat{\mathcal{L}^{-1}}\|_T$. Consequently, if $T = \mathbb{Z}$, then the existence of the sequence $(x_n)_{n \in T}$ follows directly from Theorem 92 with $\mathcal{U} = X^S$.

So, assume that $T = \mathbb{N}_0$. First consider the case where

$$p_i \in \mathbb{K}, \qquad i = 1, \ldots, m. \tag{5.91}$$

We show that hypothesis $(\mathcal{H}_i)$ (used in Theorem 89) holds for $i = 1, \ldots, m$ with

$$\rho_i(\varepsilon) = \frac{\varepsilon}{|1 - |p_i||}, \qquad \varepsilon \in \mathbb{R}_+. \tag{5.92}$$

Note that, in this situation, $(\mathcal{H}_i)$ takes the following form.

$(\mathcal{H}_i)$ For every $(z_n)_{n \in T}, (v_n)_{n \in T} \in X^T$ with

$$\varepsilon := \sup_{n \in \mathbb{N}_0} \|z_{n+1} - p_i z_n - v_n\| < \infty$$

there is a sequence $(x_n)_{n \in T} \in X^T$ such that

$$x_{n+1} - p_i x_n = v_n, \qquad n \in T, \tag{5.93}$$

and

$$\|z_n - x_n\| \leq \frac{\varepsilon}{|1 - |p_i||}, \qquad n \in T.$$

Fix $i \in \{1, \ldots, m\}$. If $|p_i| > 1$ and X is a Banach space, then $(\mathcal{H}_i)$ follows from Theorem 91 (α), with $m = 1$ and $S = T$. Therefore, assume that $|p_i| < 1$. We show the following stronger version of $(\mathcal{H}_i)$, which will be useful also for the proof of (c).

$(\widehat{\mathcal{H}_i})$ For every $\varepsilon > 0$, $u \in X$, $\|u\| \leq \varepsilon$, and $(z_n)_{n \in \mathbb{N}_0}, (v_n)_{n \in \mathbb{N}_0} \in X^{\mathbb{N}_0}$ with

$$\|z_{n+1} - p_i z_n - v_n\| \leq \varepsilon, \qquad n \in \mathbb{N}_0, \tag{5.94}$$

the sequence $(x_n(u))_{n \in \mathbb{N}_0} \in X^{\mathbb{N}_0}$, given by:

$$x_0(u) = z_0 + u, \qquad x_{n+1}(u) = p_i x_n(u) + v_n, \qquad n \in \mathbb{N}_0, \tag{5.95}$$

satisfies the condition

$$\|z_n - x_n(u)\| \leq \frac{\varepsilon}{1 - |p_i|}, \qquad n \in \mathbb{N}_0.$$

So take $\varepsilon > 0$, $u \in X$, $\|u\| \leq \varepsilon$, and $(z_n)_{n \in \mathbb{N}_0}, (v_n)_{n \in \mathbb{N}_0} \in X^{\mathbb{N}_0}$ satisfying (5.94). Write

$$b_n := z_{n+1} - p_i z_n - v_n, \qquad n \in \mathbb{N}_0.$$

Then

$$\|b_n\| \leq \varepsilon, \qquad n \in \mathbb{N}_0,$$

and

$$z_{n+1} = p_i z_n + b_n + v_n, \qquad n \in \mathbb{N}_0.$$

Next, for every $n \in \mathbb{N}$,

$$x_n(u) = p_i^n x_0(u) + \sum_{k=0}^{n-1} p_i^{n-k-1} v_k,$$

$$z_n = p_i^n z_0 + \sum_{k=0}^{n-1} p_i^{n-k-1}(b_k + v_k),$$

whence

$$\|z_n - x_n(u)\| = \left\| -p_i^n u + \sum_{k=0}^{n-1} p_i^{n-k-1} b_k \right\|$$

$$\leq \varepsilon \sum_{k=0}^{n} |p_i|^k \leq \frac{\varepsilon}{1 - |p_i|}, \qquad n \in \mathbb{N}_0.$$

Thus we have shown hypothesis $(\widehat{\mathcal{H}_i})$ is fulfilled. Consequently, Theorem 89 completes the proof when $T = \mathbb{N}_0$ and (5.91) holds.

So far we have proved the existence of the sequence $(x_n)_{n \in T}$ when (5.91) is fulfilled. It remains to consider the situation when (5.91) does not hold, i.e., $\mathbb{K} = \mathbb{R}$ and $p_i \notin \mathbb{R}$ for some $i \in \{1, \ldots, m\}$.

Let $\mathfrak{C}(X)$ be a complexification of X with the Taylor norm $\| \cdot \|_T$ and define $\pi_i : X^2 \to X$ by:

$$\pi_i(v_1, v_2) := v_i, \qquad v_1, v_2 \in X, i = 1, 2.$$

Define sequences $(\bar{y}_n)_{n \in T}$ and $(\bar{b}_n)_{n \in T}$ in X^2 by:

$$\bar{y}_n := (y_n, 0), \qquad \bar{b}_n := (b_n, 0), \qquad n \in T. \tag{5.96}$$

Then, by (5.17),

$$\|\bar{y}_{n+m} + a_{m-1}\bar{y}_{n+m-1} + \ldots + a_0\bar{y}_n - \bar{b}_n\|_T \leq \delta, \qquad n \in T. \tag{5.97}$$

Thus we have the situation when (5.91) is satisfied and consequently there exists a sequence $(\bar{x}_n)_{n \in T}$ in X^2 with

$$\bar{x}_{n+m} + a_{m-1}\bar{x}_{n+m-1} + \ldots + a_0\bar{x}_n = \bar{b}_n, \qquad n \in T, \tag{5.98}$$

$$\sup_{n \in T} \|\bar{y}_n - \bar{x}_n\|_T \leq \frac{\delta}{|1 - |p_1|| \ldots |1 - |p_m||}. \tag{5.99}$$

Let

$$x_n := \pi_1(\bar{x}_n), \qquad n \in T.$$

It is easily seen that (5.87) and (5.88) are valid (see (5.17)).

Now, we are to prove statements (a)–(c). To this end notice that, if

$$|p_i| > 1, \qquad i = 1, \ldots, m,$$

then statement (b) results directly from Proposition 7 with $S = T$ and

$$\mathcal{L}x = x \circ \xi, \qquad x \in X^T,$$

where $\xi(n) = n + 1$ for $n \in T$.

If $T = \mathbb{Z}$, then we also can use Proposition 7 with $S = T$, because then

$$\mathcal{L}^{-1}x = x \circ \xi, \qquad x \in X^T,$$

where $\xi(n) = n - 1$ for $n \in T$, and consequently $\|\mathcal{L}^{-1}\| = 1$ and $\|\widehat{\mathcal{L}^{-1}}\| = 1$.

Clearly, (b) implies the sufficient condition of (a) and the necessary condition of (a) results from (c). Consequently, it remains to prove (c).

So, assume that neither (5.90) nor (5.89) holds, i.e., $T = \mathbb{N}_0$ and $|p_j| < 1$ for some $j \in \{1, \ldots, p\}$. Without loss of generality we may assume that $|p_m| < 1$.

First observe that every sequence $(x_n)_{n \in \mathbb{N}_0}$ in X, satisfying (5.87), is uniquely determined by $x_0, \ldots, x_{m-1}$. Therefore

$$\text{card}\,\{u \in X : \|u\| \leq \varepsilon\} = \text{card}\,X = \text{card}\,X^m \qquad (5.100)$$
$$= \text{card}\,\{(x_n)_{n \in \mathbb{N}_0} \in X^{\mathbb{N}_0} : (5.87)\ \text{holds}\}.$$

Consequently the case $m = 1$ results from $(\widehat{\mathcal{H}_i})$.

Next, assume that $m > 1$. First consider the case $\mathbb{K} = \mathbb{C}$. Let $(y_n)_{n \in \mathbb{N}_0}$ be a sequence in X such that (5.86) holds. In view of Vieta's formula, (5.86) can be written in the form

$$\|y_{n+m} + (-1)(p_1 + \ldots + p_m)y_{n+m-1} + \ldots + (-1)^m(p_1 \ldots p_m)y_n - b_n\|$$
$$\leq \delta, \qquad n \in \mathbb{N}_0,$$

whence the sequence $(z_n)_{n \in \mathbb{N}_0}$, given by: $z_n := y_{n+1} - p_m y_n$, satisfies

$$\|z_{n+m-1} + (-1)(p_1 + \ldots + p_{m-1})z_{n+m-2}$$
$$+ \ldots + (-1)^{m-1}(p_1 \ldots p_{m-1})z_n - b_n\| \leq \delta, \qquad n \in \mathbb{N}_0.$$

Hence there is a sequence $(v_n)_{n \in \mathbb{N}_0}$ in X such that, for every $n \in \mathbb{N}_0$,

$$v_{n+m-1} + (-1)(p_1 + \ldots + p_{m-1})v_{n+m-2} \qquad (5.101)$$
$$+ \ldots + (-1)^{m-1}(p_1 \ldots p_{m-1})v_n = b_n,$$

$$\|z_n - v_n\| \leq \frac{\delta}{\left|1 - |p_1|\right| \ldots \left|1 - |p_{m-1}|\right|} =: \delta_0. \qquad (5.102)$$

Further

$$\|y_{n+1} - p_m y_n - v_n\| = \|z_n - v_n\| \leq \delta_0, \qquad n \in \mathbb{N}_0,$$

and, by the case $m = 1$, the cardinality of the set of all sequences $(x_n)_{n \in \mathbb{N}_0}$ in X such that

$$x_{n+1} - p_m x_n = v_n, \qquad n \in \mathbb{N}_0, \tag{5.103}$$

$$\|y_n - x_n\| \le \frac{\delta_0}{1 - |p_m|}, \qquad n \in \mathbb{N}_0, \tag{5.104}$$

is equal to the cardinality of X. Now, it is enough to notice that (5.101)–(5.104) imply (5.87) and (5.88).

Finally, assume that $\mathbb{K} = \mathbb{R}$. Let sequences $(\bar{y}_n)_{n \in \mathbb{N}_0}$ and $(\bar{b}_n)_{n \in \mathbb{N}_0}$ in X^2 be given by:

$$\bar{y}_n := (y_n, 0), \qquad \bar{b}_n := (b_n, 0), \qquad n \in \mathbb{N}_0.$$

Then

$$\|\bar{y}_{n+m} + a_{m-1}\bar{y}_{n+m-1} + \ldots + a_0\bar{y}_n - \bar{b}_n\|_T \le \delta, \qquad n \in \mathbb{N}_0.$$

Given $u \in X$, we define a sequence $(\bar{x}_n((u, 0)))_{n \in \mathbb{N}_0}$ in X^2 by:

$$\bar{x}_0((u, 0)) := \bar{y}_0 + (u, 0), \qquad \bar{x}_{n+1}((u, 0)) = p_m \bar{x}_n((u, 0)) + \bar{v}_n, \qquad n \in \mathbb{N}_0.$$

Clearly, $(\widehat{\mathcal{H}_i})$ implies that, if $\|u\| \le \delta$, then

$$\|\bar{y}_n - \bar{x}_n((u, 0))\|_T \le \frac{\delta_0}{1 - |p_m|}, \qquad n \in \mathbb{N}_0, \tag{5.105}$$

whence

$$\|y_n - \pi_1(\bar{x}_n((u, 0)))\| \le \frac{\delta}{\left|1 - |p_1|\right| \ldots \left|1 - |p_m|\right|}, \qquad n \in \mathbb{N}_0.$$

Moreover, the sequence $(\pi_1(\bar{x}_n((u, 0))))_{n \in \mathbb{N}_0}$ satisfies recurrence (5.87) with each $u \in X$ and, for every $u_1, u_2 \in X$ with $u_1 \ne u_2$,

$$\pi_1(\bar{x}_0((u_1, 0))) = y_0 + u_1 \ne y_0 + u_2 = \pi_1(\bar{x}_0((u_2, 0))).$$

This and (5.100) yield (c) and therefore the necessary condition of (a).   □

Remark 22. Clearly,

$$z^m + \sum_{i=0}^{m-1} a_i z^i = \prod_{i=1}^{m} (z - p_i), \qquad z \in \mathbb{C},$$

whence (with $z = 1$) we get

$$1 + \sum_{i=0}^{m-1} a_i = \prod_{i=1}^{m} (1 - p_i). \tag{5.106}$$

Assume that $T \in \{\mathbb{N}_0, \mathbb{Z}\}$, $p_1, \ldots, p_m \in \mathbb{R} \setminus \{1\}$, and $b \in X$. According to (5.106),

$$\beta := 1 + \sum_{i=0}^{m-1} a_i \neq 0.$$

Define sequences $(x_n)_{n \in T}$ and $(y_n)_{n \in T}$ in X by the following:

$$y_n := 0, \qquad x_n := \frac{1}{\beta} b, \qquad n \in T,$$

with a fixed $b \in X$. Then

$$\|y_{n+m} + a_{m-1} y_{n+m-1} + \ldots + a_0 y_n - b\| = \|b\|, \qquad n \in T,$$

$$x_{n+m} = \frac{1}{\beta} b = (1 - \beta) \frac{1}{\beta} b + b$$

$$= -a_{m-1} x_{n+m-1} - \ldots - a_0 x_n + b, \qquad n \in T,$$

$$\|y_n - x_n\| = \frac{\|b\|}{|\beta|} = \frac{\|b\|}{|1 - p_1| \ldots |1 - p_m|}, \qquad n \in T. \tag{5.107}$$

So, in view of the uniqueness of $(x_n)_{n \in T}$ in Theorem 95, we obtain the conclusion that estimation (5.88) is optimum when either (5.89) or (5.90) holds, and $p_i \geq 0$ for $i = 1, \ldots, m$.

Remark 23 shows that this is not the case in the general situation (cf. [4]).

We have the following example:

Example 14. The following difference equation is an easy and well known example of (5.87):

$$x_{n+2} = x_{n+1} + x_n, \qquad n \in \mathbb{N}_0.$$

If $x_0 = 0$ and $x_1 = 1$, then it defines the Fibonacci sequence:

$$0, 1, 1, 2, 3, 5, 8, 13, 21, 34, \ldots$$

The characteristic equation of it is

$$z^2 = z + 1$$

with the roots

$$p_1 = \frac{1 + \sqrt{5}}{2}, \qquad p_2 = \frac{1 - \sqrt{5}}{2}.$$

Clearly $|p_1| \neq 1 \neq |p_2|$ and (5.88) takes the form

$$|y_n - x_n| \leq \delta(2 + \sqrt{5}), \qquad n \in \mathbb{N}_0. \tag{5.108}$$

Theorem 96. *Let X be a Banach space, $T \in \{\mathbb{N}_0, \mathbb{Z}\}$, $\delta > 0$, and $(y_n)_{n \in T}$ and $(b_n)_{n \in T}$ be two sequences in X such that*

$$\|y_{n+2} + a_1 y_{n+1} + a_0 y_n - b_n\| \leq \delta, \qquad n \in T. \tag{5.109}$$

Assume that

$$p_1 \neq p_2, \qquad p_1 p_2 \neq 0, \tag{5.110}$$

and one of the following two hypotheses is valid.
(α) $|p_i| > 1$ for $i = 1, 2$.
(β) $T = \mathbb{Z}$ and $|p_i| \neq 1$ for $i = 1, 2$.
Then there exists a unique sequence $(x_n)_{n \in T}$ in X such that

$$x_{n+2} + a_1 x_{n+1} + a_0 x_n = b_n, \qquad n \in T, \tag{5.111}$$

and

$$\|y_n - x_n\| \leq \frac{\delta}{|p_2 - p_1|} \left(\frac{1}{|1 - |p_1||} + \frac{1}{|1 - |p_2||} \right), \qquad n \in T. \tag{5.112}$$

Proof. The existence of a sequence $(x_n)_{n \in T}$ follows directly from Theorems 93 and 94, with $\mathcal{U} = X^T$ and

$$\mathcal{L}(x) = x \circ \xi, \qquad x \in X^T,$$

with $\xi(n) = n + 1$ for $n \in T$ (clearly, if $T = \mathbb{Z}$, then $\mathcal{L}$ is bijective). It is only enough to observe that

$$\|\mathcal{L}\| = 1 = \|\widehat{\mathcal{L}}\|_T.$$

Finally, observe that the statement on uniqueness of $(x_n)_{n \in T}$ is a consequence of Theorem 95 (b). $\qquad \square$

It is easily seen that Theorems 95 and 96 imply the following.

Corollary 8. *Let X be a Banach space, $T \in \{\mathbb{N}_0, \mathbb{Z}\}$, $\delta > 0$, and $(y_n)_{n \in T}$ and $(b_n)_{n \in T}$ be two sequences in X satisfying (5.109). Assume that (5.110) holds and one of the following two hypotheses is valid:*
(α) $|p_i| > 1$ for $i = 1, 2$.
(β) $T = \mathbb{Z}$ and $|p_i| \neq 1$ for $i = 1, 2$.

Then there exists a unique sequence $(x_n)_{n \in T}$ in X such that (5.111) is valid and

$$\|y_n - x_n\| \le \mu \delta, \tag{5.113}$$

where

$$\mu = \min\{\mu_1, \mu_2\}$$

and

$$\mu_1 := \frac{1}{|p_1 - p_2|}\Big(\frac{1}{\|p_1\| - 1|} + \frac{1}{\|p_2\| - 1|}\Big),$$

$$\mu_2 := \frac{1}{|(|p_1| - 1)(|p_2| - 1)|}.$$

Remark 23. Note that, if $p_1 = 2$ and $p_2 = -2$, then

$$\mu_1 = \frac{1}{2} < \mu_2 = 1.$$

Moreover, in this case, estimate (5.107) in Remark 22, with $m = 2$ and $\|b\| = \delta$, takes the form

$$\|y_n - x_n\| = \frac{\delta}{|1 - p_1| |1 - p_2|} = \frac{\delta}{3}, \qquad n \in T.$$

But with $p_1 = -1/4$ and $p_2 = 1/4$ we get

$$\mu_1 = 16/3 > \mu_2 = 16/9.$$

This shows that neither (5.88) nor (5.112) are optimum in general for $m = 2$ and $p_1 \ne p_2$. In this way there arises a natural problem to find such generally optimal estimate.

6. Difference equation with a matrix coefficient

Let Z be a Banach space over $\mathbb{K}$. Now we present some results concerning the Hyers-Ulam stability of the following system of first-order linear difference equations, with constant coefficients $a_{ij} \in \mathbb{K}$, $i, j = 1, \ldots, r$ ($r \in \mathbb{N}$ is fixed):

$$\begin{cases} x_{n+1}^1 = a_{11}x_n^1 + a_{12}x_n^2 + \cdots + a_{1r}x_n^r + d_n^1, \\[2mm] x_{n+1}^2 = a_{21}x_n^1 + a_{22}x_n^2 + \cdots + a_{2r}x_n^r + d_n^2, \\[2mm] \cdots \quad \cdots \qquad \cdots \quad \cdots \\[2mm] x_{n+1}^r = a_{r1}x_n^1 + a_{r2}x_n^2 + \cdots + a_{rr}x_n^r + d_n^r, \end{cases} \tag{5.114}$$

for sequences $(x_n)_{n \in \mathbb{N}_0}$ in Z^r with $x_n := (x_n^1, \ldots, x_n^r)$ for all $n \in \mathbb{N}_0$, where $d_n^1, \ldots, d_n^r \in Z$ for $n \in \mathbb{N}_0$ are fixed. They have been motivated by the results in [32]. Clearly, those results somehow complement the theorems from the previous section.

Note that if

$$
A = \begin{bmatrix} a_{11} & a_{12} & \cdots & a_{1r} \\ a_{21} & a_{22} & \cdots & a_{2r} \\ \vdots & \vdots & \ddots & \vdots \\ a_{r1} & a_{r2} & \cdots & a_{rr} \end{bmatrix}, \qquad \mathbf{x}_n = \begin{bmatrix} x_n^1 \\ x_n^2 \\ \vdots \\ x_n^r \end{bmatrix}, \qquad \mathbf{d}_n = \begin{bmatrix} d_n^1 \\ d_n^2 \\ \vdots \\ d_n^r \end{bmatrix},
$$

then (5.114) can be expressed in the following simple form

$$
\mathbf{x}_{n+1} = A\mathbf{x}_n + \mathbf{d}_n, \qquad n \in \mathbb{N}_0. \tag{5.115}
$$

Next, as in [32], to simplify the notations, we consider $\mathbf{x}_n$ and $\mathbf{d}_n$ to be elements of Z^r, when it is convenient (and makes no confusion); i.e., we identify $\mathbf{x}_n$ with $(x_n^1, \ldots, x_n^r)$ and $\mathbf{d}_n$ with $(d_n^1, \ldots, d_n^r)$.

Further, for each $\mathbf{x} = (x^1, \ldots, x^r) \in Z^r$ and any matrix

$$
B = \begin{bmatrix} b_{11} & b_{12} & \cdots & b_{1r} \\ b_{21} & b_{22} & \cdots & b_{2r} \\ \vdots & \vdots & \ddots & \vdots \\ b_{r1} & b_{r2} & \cdots & b_{rr} \end{bmatrix} \tag{5.116}
$$

we write

$$
\|B\| := \max_{1 \le i \le r} \sum_{j=1}^{r} |b_{ij}|,
$$

$$
\|\mathbf{x}\|_m := \max_{1 \le i \le r} \|x^i\|.
$$

Clearly, since Z is a Banach space, then so is Z^r, when endowed with the norm $\| \cdot \|_m$.

In what follows we assume that A is nonsingular and $\lambda_1, \ldots, \lambda_l$ are the eigenvalues of the matrix A^{-1} with multiplicities $r_1, \ldots, r_l$, respectively. Clearly, if $i \in \{1, \ldots, m\}$ and $p_i \ne 0$, then $p_i\lambda_1, \ldots, p_i\lambda_l$ are the eigenvalues of the matrix $A_i := p_iA^{-1}$ also with multiplicities $r_1, \ldots, r_l$, respectively. Hence, for each $i \in \{1, \ldots, m\}$ with $p_i \ne 0$, there exists a nonsingular matrix Q_i in $\mathbb{C}^{r \times r}$ with

$$
A_i = Q_iJ_iQ_i^{-1},
$$

where

$$
J_i = J_i^{\lambda_1, r_1} \oplus \ldots \oplus J_i^{\lambda_l, r_l},
$$

$$
J_i^{\lambda_j, r_j} =
\begin{bmatrix}
p_i \lambda_j & 1 & 0 & \cdots & 0 & 0 \\
0 & p_i \lambda_j & 1 & \cdots & 0 & 0 \\
\vdots & \vdots & \vdots & \ddots & \vdots & \vdots \\
0 & 0 & 0 & \cdots & p_i \lambda_j & 1 \\
0 & 0 & 0 & \cdots & 0 & p_i \lambda_j
\end{bmatrix}_{r_j \times r_j}
, \qquad j = 1, \ldots, l.
$$

If $p_i = 0$, then we take as Q_i the identity (unit) matrix I_r in $\mathbb{C}^{r \times r}$.

Now, we are in a position to present a proposition that easily follows from the main result in [32, Theorem 5].

Proposition 8. *Assume that $i \in \{1, \ldots, m\}$ and $|p_i \lambda_j| \neq 1$ for $j = 1, \ldots, l$. For any sequences $(\mathbf{z}_n)_{n \in \mathbb{N}_0}$ and $(\widetilde{\mathbf{d}}_n)_{n \in \mathbb{N}_0}$ in Z^r, satisfying*

$$
\delta_0 := \sup_{n \in \mathbb{N}_0} \|\mathbf{z}_{n+1} + A_i \mathbf{z}_n - \widetilde{\mathbf{d}}_n\|_m < \infty, \tag{5.117}
$$

there exists a sequence $(\mathbf{x}_n)_{n \in \mathbb{N}_0}$ in Z^r such that

$$
\mathbf{x}_{n+1} = -A_i \mathbf{x}_n + \widetilde{\mathbf{d}}_n, \qquad n \in \mathbb{N}_0, \tag{5.118}
$$

$$
\sup_{n \in \mathbb{N}_0} \|\mathbf{z}_n - \mathbf{x}_n\|_m \leq \delta_0 \|Q_i\| \|Q_i^{-1}\| \max_{j=1,\ldots,l} \sum_{k=1}^{r_j} \frac{1}{\left|1 - |p_i \lambda_j|\right|^k}. \tag{5.119}
$$

Proposition 8 yields the following:

Corollary 9. *Assume that $i \in \{1, \ldots, m\}$ and $|p_i \lambda_j| \neq 1$ for $j = 1, \ldots, l$. For any sequence $(\mathbf{w}_n)_{n \in \mathbb{N}_0}$ in Z^r, satisfying*

$$
\delta := \sup_{n \in \mathbb{N}_0} \|A \mathbf{w}_{n+1} + p_i \mathbf{w}_n - \mathbf{d}_n\|_m < \infty, \tag{5.120}
$$

there exists a sequence $(\mathbf{y}_n)_{n \in \mathbb{N}_0}$ in Z^r such that

$$
A \mathbf{y}_{n+1} + p_i \mathbf{y}_n = \mathbf{d}_n, \qquad n \in \mathbb{N}_0, \tag{5.121}
$$

$$
\sup_{n \in \mathbb{N}_0} \|\mathbf{w}_n - \mathbf{y}_n\|_m \leq \delta \|Q_i\| \|Q_i^{-1}\| \|A^{-1}\| \max_{j=1,\ldots,l} \sum_{k=1}^{r_j} \frac{1}{\left|1 - |p_i \lambda_j|\right|^k}. \tag{5.122}
$$

Proof. Let $(\mathbf{w}_n)_{n \in \mathbb{N}_0}$ be a sequence in Z^r satisfying (5.120). If $p_i = 0$, then it is enough to take $\mathbf{y}_n = A^{-1} \mathbf{d}_n$ for $n \in \mathbb{N}_0$. Then

$$
\|\mathbf{w}_n - \mathbf{y}_n\|_m = \|\mathbf{w}_n - A^{-1} \mathbf{d}_n\|_m \leq \|A^{-1}\| \|A \mathbf{w}_n - \mathbf{d}_n\|_m
$$
$$
\leq \delta \|A^{-1}\|, \qquad n \in \mathbb{N}_0,
$$

which implies (5.122), because in this case (by definition) we take $Q_i := I_r$ (the identity matrix) and therefore $\|Q_i\| \|Q_i^{-1}\| = 1$.

Now, assume that $p_i \neq 0$. By (5.120), we have

$$
\begin{aligned}
\delta_0 : \;=\; & \sup_{n \in \mathbb{N}_0} \|\mathbf{w}_{n+1} + A_i \mathbf{w}_n - \widetilde{\mathbf{d}}_n\|_m \\
\leq \;& \|A^{-1}\| \sup_{n \in \mathbb{N}_0} \|A\mathbf{w}_{n+1} + p_i \mathbf{w}_n - \mathbf{d}_n\|_m \\
\leq \;& \|A^{-1}\|\delta,
\end{aligned}
$$

where $\widetilde{\mathbf{d}}_n := A^{-1}\mathbf{d}_n$ for $n \in \mathbb{N}_0$. By Proposition 8, there exists a sequence $(\mathbf{y}_n)_{n \in \mathbb{N}_0}$ in Z^r such that

$$
\mathbf{y}_{n+1} = -A_i \mathbf{y}_n + \widetilde{\mathbf{d}}_n, \qquad n \in \mathbb{N}_0,
$$

$$
\sup_{n \in \mathbb{N}_0} \|\mathbf{w}_n - \mathbf{y}_n\|_m \leq \delta_0 \|Q_i\| \|Q_i^{-1}\| \max_{j=1,\dots,l} \sum_{k=1}^{r_j} \frac{1}{\left|1 - |p_i \lambda_j|\right|^k}.
$$

Clearly, (5.121) and (5.122) hold.   □

Now, from our previous outcomes we can deduce, for instance, the following three theorems on stability of the difference equation

$$
A^m \mathbf{x}_{n+m} + \sum_{i=0}^{m-1} a_i A^i \mathbf{x}_{n+i} = \mathbf{d}_n, \qquad n \in \mathbb{N}_0. \tag{5.123}
$$

Theorem 97. *Assume that*

$$
p_i \in \mathbb{K}, \qquad i = 1, \dots, m, \tag{5.124}
$$

and $|p_i \lambda_j| \neq 1$ for $i = 1, \dots, m$, $j = 1, \dots, l$. Then, for any sequence $(\mathbf{z}_n)_{n \in \mathbb{N}_0}$ in Z^r, satisfying

$$
\delta := \sup_{n \in \mathbb{N}_0} \left\| A^m \mathbf{z}_{n+m} + \sum_{i=0}^{m-1} a_i A^i \mathbf{z}_{n+i} - \mathbf{d}_n \right\|_m < \infty, \tag{5.125}
$$

there exists a sequence $(\mathbf{x}_n)_{n \in \mathbb{N}_0}$ in Z^r such that

$$
A^m \mathbf{x}_{n+m} + \sum_{i=0}^{m-1} a_i A^i \mathbf{x}_{n+i} = \mathbf{d}_n, \qquad n \in \mathbb{N}_0, \tag{5.126}
$$

$$
\sup_{n \in \mathbb{N}_0} \|\mathbf{z}_n - \mathbf{x}_n\|_m \leq \delta \omega_1 \dots \omega_m, \tag{5.127}
$$

where

$$\omega_i := \|Q_i\| \, \|Q_i^{-1}\| \, \|A^{-1}\| \, \max_{j=1,\ldots,l} \sum_{k=1}^{r_j} \frac{1}{\left|1 - |p_i \lambda_j|\right|^k}, \qquad i = 1, \ldots, m. \tag{5.128}$$

Proof. It is easily seen that, by Corollary 8, hypothesis $(\mathcal{H}_i)$ holds for $i = 1, \ldots, m$ with $\rho_i(\epsilon) = \omega_i \epsilon$ for $\epsilon \in \mathbb{R}_+$. So, it is enough to use Theorem 89 with $S = \mathbb{N}_0$, $X = Z^r$, $\mathcal{U} = X^S$, and

$$\mathcal{L}(\mathbf{z}_n)_{n \in \mathbb{N}_0} = (A\mathbf{z}_{n+1})_{n \in \mathbb{N}_0}, \qquad (\mathbf{z}_n)_{n \in \mathbb{N}_0} \in X^S, \tag{5.129}$$

$$F(n) = \mathbf{d}_n, \qquad n \in S. \tag{5.130}$$

$\square$

Theorem 91 yields an estimation somewhat different from (5.127), but valid also in the case where (5.124) does not hold.

Theorem 98. *Let*

$$\kappa := \begin{cases} 1, & \text{if (5.124) holds;} \\ \sqrt{2}, & \text{otherwise,} \end{cases} \tag{5.131}$$

and $|p_i| > \kappa \|A\|$ for $i = 1, \ldots, m$. Then, for any sequence $(\mathbf{z}_n)_{n \in \mathbb{N}_0}$ in Z^r, satisfying (5.125), there exists a unique sequence $(\mathbf{x}_n)_{n \in \mathbb{N}_0}$ in Z^r such that (5.126) holds and

$$\sup_{n \in \mathbb{N}_0} \|\mathbf{z}_n - \mathbf{x}_n\|_m \leq \frac{\delta}{(|p_1| - \kappa \|A\|) \ldots (|p_m| - \kappa \|A\|)}. \tag{5.132}$$

Proof. It is enough to use Proposition 5 and Theorem 91 with $S = \mathbb{N}_0$, $X = Z^r$, $\mathcal{U} = X^S$, and $\mathcal{L}$ and F given by (5.129) and (5.130), because $\|\mathcal{L}\| = \|A\|$. $\square$

The next theorem follows from Theorems 93 and 94.

Theorem 99. *Assume that (5.44) and (5.46) are valid and $|p_i| > \kappa \|A\| =: L$ for $i = 1, 2$, where κ is given by (5.131). Then, for any sequence $(\mathbf{z}_n)_{n \in \mathbb{N}_0}$ in Z^r, satisfying*

$$\delta := \sup_{n \in \mathbb{N}_0} \left\| A^2 \mathbf{z}_{n+2} + a_1 A \mathbf{z}_{n+1} + a_0 \mathbf{z}_n - \mathbf{d}_n \right\|_m < \infty, \tag{5.133}$$

there exists a unique sequence $(\mathbf{x}_n)_{n \in \mathbb{N}_0}$ in Z^r such that

$$A^2 \mathbf{x}_{n+2} + a_1 A \mathbf{x}_{n+1} + a_0 \mathbf{x}_n = \mathbf{d}_n, \qquad n \in \mathbb{N}_0, \tag{5.134}$$

$$\sup_{n\in\mathbb{N}_0} \|\mathbf{z}_n - \mathbf{x}_n\|_m \le \frac{\delta}{|p_2 - p_1|}\left(\frac{1}{|L - |p_1||} + \frac{1}{|L - |p_2||}\right). \qquad (5.135)$$

Proof. It is enough to use Proposition 5 and Theorems 93 and 94, with $S = \mathbb{N}_0$, $X = Z^r$, $\mathcal{U} = X^S$, and $\mathcal{L}$ and F given by (5.129) and (5.130). $\qquad\square$

7. Linear functional equations with constant coefficients

In a short and simple way it has been shown in [8] that, in some cases, the Hyers-Ulam stability of the functional equation

$$\varphi(f(x)) = a\varphi(x) + G(x), \qquad x \in S, \qquad (5.136)$$

(in the class of functions $\varphi : S \to X$ and with $a \in \mathbb{K}$, $f : S \to S$ and $G : S \to X$ given) implies the Hyers-Ulam stability of the equation

$$\varphi(f^m(x)) + \sum_{j=0}^{m-1} a_j\varphi(f^j(x)) = F(x), \qquad x \in S, \qquad (5.137)$$

for every $m \in \mathbb{N}$. In particular, the following result can be easily derived from [8].

Theorem 100. *Suppose that $\delta \in \mathbb{R}_+$, $f : S \to S$ is bijective, $\varphi_s : S \to X$ satisfies*

$$\left\|\varphi_s(f^m(x)) + \sum_{j=0}^{m-1} a_j\varphi_s(f^j(x)) - F(x)\right\| \le \delta, \qquad x \in S, \qquad (5.138)$$

and $|p_i| \ne 1$ for $i = 1, \ldots, m$. Then there exists a solution $\varphi : S \to X$ of equation (5.137) *such that*

$$\|\varphi_s(x) - \varphi(x)\| \le \frac{\delta}{|1 - |p_1||\ldots|1 - |p_m||}, \qquad x \in S.$$

It is easily seen that Theorem 100 can also be deduced from our Theorem 91 (with $\mathcal{U} = X^S$ and $\mathcal{L}\phi = \phi \circ f$). Now, we show that analogous results follow from Theorem 91 for a somewhat more general equation

$$A^m(\varphi(f^m(x))) + \sum_{j=0}^{m-1} a_jA^j(\varphi(f^j(x))) = F(x), \qquad x \in S, \qquad (5.139)$$

where linear $A : X \to X$ is given.

To this end we need the following notion:

$$[A] := \sup\{t \in \mathbb{R}_+ : \|A(x)\| \ge t\|x\| \text{ for } x \in X\}.$$

Clearly, if $A(x) = cx$ for every $x \in X$, with some $c \in \mathbb{K}$, then

$$\|A\| = [A] = |c|;$$

if $c = 1$, then (5.139) becomes (5.137).

Remark 24. It is easily seen that if A is bijective, then

$$[A] = \frac{1}{\|A^{-1}\|}.$$

Let us start with the following auxiliary proposition.

Proposition 9. *Assume that A is bounded, κ is given by (5.131), and one of the following two conditions is valid:*
(A) $|p_i| > \kappa\|A\|$ *for $i = 1, \ldots, m$.*
(B) f *is surjective and, for each $i \in \{1, \ldots, m\}$,*

$$\kappa|p_i| < [A] \quad or \quad |p_i| > \kappa\|A\|.$$

Then, for every $F \in X^S$ and $\gamma \in X^S$, there exists at most one solution $\varphi \in X^S$ of equation (5.139) such that

$$\|\gamma - \varphi\|^* < \infty.$$

Proof. The case

$$|p_i| > \kappa\|A\|, \qquad i = 1, \ldots, m,$$

can be easily derived from Proposition 7, but for the convenience of readers we present the proof also in that case (which makes the reasonings more natural).

First consider the case $\mathbb{K} = \mathbb{C}$. We show, by induction on m, that if $\psi, \psi' \in X^S$ are solutions of (5.139) such that

$$\|\psi - \psi'\|^* =: M < \infty,$$

then $\psi = \psi'$. For $m = 1$ we have $a_0 = -p_1$ and

$$|a_0|^n\|\psi(x) - \psi'(x)\| = \|A^n(\psi(f^n(x)) - \psi'(f^n(x)))\| \le \|A\|^n M$$

for $n \in \mathbb{N}$, $x \in S$, whence $\psi = \psi'$ in the case $|p_1| > \|A\|$. If f is surjective, then for each $n \in \mathbb{N}$ and $x \in S$ there is $x_n \in S$ with $f^n(x_n) = x$ and, consequently,

$$[A]^n\|\psi(x) - \psi'(x)\| \le \|A^n(\psi(x) - \psi'(x))\|$$
$$= |a_0|^n\|\psi(x_n) - \psi'(x_n)\| \le |a_0|^n M,$$

which also yields $\psi = \psi'$ when $|p_1| < [A]$.

So, fix now $k \in \mathbb{N}$ and assume that the inductive statement is true for $m = k$. We are to show that this is also the case for $m = k + 1$. To this end take solutions $\psi, \psi' \in X^S$ of (5.139) such that

$$M := \|\psi - \psi'\|^* < \infty.$$

Write

$$\eta(x) := A(\psi(f(x))) - p_{k+1}\psi(x), \qquad x \in S,$$

$$\eta'(x) := A(\psi'(f(x))) - p_{k+1}\psi'(x), \qquad x \in S.$$

Then $\eta, \eta' \in X^S$ are solutions of (5.139) with $m = k$ and $a_0, \ldots, a_{k-1}$ replaced by $b_0, \ldots, b_{k-1}$, described in Remark 19. Moreover,

$$\|\eta - \eta'\|^* \le (\|A\| + |p_{k+1}|)M.$$

Consequently, in view of the inductive hypothesis, $\eta = \eta'$ and consequently, arguing analogously as in the case $m = 1$, we get $\psi = \psi'$.

Now, to complete the proof consider the case $\mathbb{K} = \mathbb{R}$. Fix $F, \gamma \in X^S$. Let $\varphi_1, \varphi_2 \in X^S$ be solutions of (5.139) such that

$$\|\gamma - \varphi_i\|^* < \infty, \qquad i = 1, 2.$$

Then

$$\|\varphi_1 - \varphi_2\|^* < \infty.$$

Write

$$F_0(x) := (F(x), 0), \qquad \widehat{\varphi_i}(x) := (\varphi_i(x), 0), \qquad x \in S, i = 1, 2,$$

$$\widehat{A}(x, y) = (A(x), A(y)), \qquad x, y \in X.$$

Then $\widehat{A}$ is a $\mathbb{C}$-linear endomorphism of $\mathfrak{C}(X)$ (the complexification of X). Next, note that $\widehat{\varphi_1}, \widehat{\varphi_2}$ are solutions of (5.139) (with F and A replaced by F_0 and $\widehat{A}$) and

$$\|\widehat{\varphi_1} - \widehat{\varphi_2}\|_T^* < \infty.$$

Hence, by the first part of the proof, $\widehat{\varphi_1} = \widehat{\varphi_2}$ and consequently $\varphi_1 = \varphi_2$. □

Now we have all tools to prove the subsequent theorem.

Theorem 101. *Suppose that $\delta \in \mathbb{R}_+$, $\varphi_s : S \to X$ satisfies*

$$\left\| A^m(\varphi_s(f^m(x))) + \sum_{j=0}^{m-1} a_j A^j(\varphi_s(f^j(x))) - F(x) \right\| \le \delta, \qquad x \in S, \tag{5.140}$$

and one of the following three hypotheses is valid.

(i) $|p_i| > \|A\|$ *for* $i = 1, \ldots, m$ *and* (5.124) *holds.*

(ii) A *and* f *are bijective,* $[A] > 0$, (5.124) *holds and* $|p_i| \notin [[A], \|A\|]$ *for* $i = 1, \ldots, m$
(if $\|A\| = \infty$, *then the condition* $|p_i| \notin [[A], \|A\|]$ *simply means that* $|p_i| < [A]$).

(iii) $|p_i| > \sqrt{2} \, \|A\|$ *for* $i = 1, \ldots, m$.

(iv) A *and* f *are bijective,* $[A] > 0$ *and* $|p_i| \notin [[A]/\sqrt{2}, \sqrt{2}\|A\|]$ *for* $i = 1, \ldots, m$.

Then there exists a unique solution $\varphi : S \to X$ *of equation* (5.139) *such that*

$$\|\varphi_s - \varphi\|^* \leq \frac{\delta}{\big||p_1| - \rho_1\big| \ldots \big||p_m| - \rho_i\big|}, \tag{5.141}$$

where

$$\rho_i := \begin{cases} \|A\|, & \text{if } p_i > \|A\| \text{ and } (5.124) \text{ holds;} \\ [A], & \text{if } p_i < [A] \text{ and } (5.124) \text{ holds;} \\ \sqrt{2}\,\|A\|, & \text{if } p_i > \sqrt{2}\|A\|; \\ [A]/\sqrt{2}, & \text{if } \sqrt{2}p_i < [A]. \end{cases}$$

Proof. The cases of (*i*) and (*iii*) follow directly from Proposition 5 and Theorem 91 with $\mathcal{U} = X^S$ and $\mathcal{L}\phi \equiv A \circ \phi \circ f$ (see Remark 24). If (*ii*) and (*iv*) holds, then the statement results analogously from Theorem 92. So, Proposition 9 completes the proof.   □

If $\sigma : S \to \mathbb{K}$, $f : S \to S$, and $\mathcal{L}g := \sigma g \circ f$ for $g \in \mathcal{U}$, then

$$\|\mathcal{L}\| = \sup_{t \in S} |\sigma(t)| \tag{5.142}$$

and equation (5.5) takes the form

$$\sigma_m(t)\varphi(f^m(t)) + \sum_{j=0}^{m-1} a_j \sigma_j(t)\varphi(f^j(t)) = F(t), \tag{5.143}$$

where $\sigma_0(t) = 1$ and $\sigma_j(t) = \sigma_{j-1}(t)\sigma(f^{j-1}(t))$ for $t \in S$, $j = 1, \ldots, m$. The form of σ_j seems to be a bit complicated for greater m, but for instance with $m = 2$ equation (5.143) has the following simple and quite general form

$$\sigma(t)\sigma(f(t))\varphi(f^2(t)) + a_1\sigma(t)\varphi(f(t)) + a_0\varphi(t) = F(t). \tag{5.144}$$

From Theorem 93 we can derive in particular the following result concerning the homogeneous version of (5.144), i.e., of the functional equation

$$\sigma(t)\sigma(f(t))\varphi(f^2(t)) + a_1\sigma(t)\varphi(f(t)) + a_0\varphi(t) = 0. \tag{5.145}$$

Theorem 102. *Let $\varepsilon > 0$ and $\varphi_s : S \to X$ satisfy the inequality*

$$\sup_{x \in S} \|\sigma(t)\sigma(f(t))\varphi_s(f^2(x)) + a_1\sigma(t)\varphi_s(f(x)) + a_0\varphi_s(x)\| \le \varepsilon. \tag{5.146}$$

Let

$$\iota := \inf_{t \in S} |\sigma(t)|, \qquad s := \sup_{t \in S} |\sigma(t)|.$$

Suppose that $p_1 \ne p_2$, $a_0 \ne 0$ and one of the following two conditions is valid:

(α) $|p_i| > s$ *for $i = 1, 2$.*

(β) f *is bijective, $\iota > 0$, and $|p_i| \notin [\iota, s]$ for $i = 1, 2$ (if $s = \infty$, then the condition $|p_i| \notin [\iota, s]$ simply means that $|p_i| < \iota$).*

Then there exists a unique solution $\varphi : S \to X$ of equation (5.145) such that

$$\sup_{x \in S} \|\varphi_s(x) - \varphi(x)\| \le \frac{\varepsilon}{|p_1 - p_2|}\left(\frac{1}{||p_1| - \mu_1|} + \frac{1}{||p_2| - \mu_2|}\right), \tag{5.147}$$

where

$$\mu_i := \begin{cases} s, & \text{if } |p_i| > s; \\ \iota, & \text{if } |p_i| < \iota. \end{cases}$$

Proof. Note that, if $\mathcal{L}g := \sigma g \circ f$ for $g \in X^S$, then $\|\mathcal{L}\| = s$ (see (5.142)). Further, if $\iota > 0$ and f is bijective, then $\mathcal{L}$ is bijective,

$$\mathcal{L}^{-1}g(x) = \frac{1}{\sigma(x)}g \circ f^{-1}(x), \qquad g \in X^S, x \in S,$$

and consequently, if $\iota > 0$, then $\|\mathcal{L}^{-1}\| = 1/\iota$. Analogously we obtain that $\|\widehat{\mathcal{L}}\|_T = s$ and $\|\widehat{\mathcal{L}}^{-1}\|_T = 1/\iota$ (when $\iota > 0$).

Now it is enough to use Theorems 93 and 94. □

From Theorem 102 and Proposition 4, we can easily derive stability results also for (5.144).

Finally, consider the homogeneous version of the linear functional equation (5.137) of order 2, i.e., the equation

$$\varphi(f^2(x)) + a_1\varphi(f(x)) + a_0\varphi(x) = 0. \tag{5.148}$$

Let us observe that Theorem 100 yields the following:

Corollary 10. *Let $|p_i| \ne 1$ for $i = 1, 2$, f be bijective, and $\varphi_s : S \to X$ satisfy the condition*

$$\varepsilon := \sup_{x \in S} \|\varphi_s(f^2(x)) + a_1\varphi_s(f(x)) + a_0\varphi_s(x)\| < \infty. \tag{5.149}$$

Then there is a unique solution $\varphi : S \to X$ *of* (5.148) *with*

$$\sup_{x \in S} \|\varphi_s(x) - \varphi(x)\| \le \frac{\varepsilon}{|(|p_1| - 1)(|p_2| - 1)|}. \tag{5.150}$$

Remark 25. Analogously as in the case of difference equations, for bijective f, Theorem 102 (with $\sigma(x) \equiv 1$) improves estimation (5.150) in some cases; however, in some other situations estimation (5.150) is better (see Remark 23). Note yet that, in Theorem 102, f can be quite arbitrary in the case of (α). So we can state the following.

Corollary 11. *Let* $\varepsilon > 0$ *and* $\varphi : S \to X$ *satisfy* (5.149). *Suppose that one of the following three conditions is valid:*

(i) $|p_i| > 1$ *for* $i = 1, 2$, $a_0 \ne 0$ *and* $p_1 \ne p_2$.
(ii) $|p_i| \ne 1$ *for* $i = 1, 2$ *and* f *is bijective.*
(iii) $a_0 \ne 0$, $p_1 \ne p_2$ *and* (ii) *holds.*

Then there exists a solution $\varphi : S \to X$ *of equation* (5.148) *such that*

$$\sup_{x \in S} \|\varphi_s(x) - \varphi(x)\| \le M\varepsilon, \tag{5.151}$$

where

$$M = \begin{cases} \min\{M_1, M_2\} & \text{if (i) or (iii) holds;} \\ M_2 & \text{if (ii) holds} \end{cases}$$

and

$$M_1 := \frac{1}{|p_1 - p_2|}\left(\frac{1}{\|p_1\| - 1|} + \frac{1}{\|p_2\| - 1|}\right),$$

$$M_2 := \frac{1}{|(|p_1| - 1)(|p_2| - 1)|}.$$

Using Proposition 4, we can easily deduce from Corollary 11 analogous stability results also for the equation

$$\varphi(f^2(x)) + a_1\varphi(f(x)) + a_0\varphi(x) = F(x). \tag{5.152}$$

The next result, which can be easily derived from [20, Theorem 3.1], shows that in some particular cases other sometimes better estimations are possible, as well.

Theorem 103. *Let* $a_1, a_0 \in \mathbb{R}$, $a_1^2 - 4a_0 \ne 0$, $0 < |p_2| < 1 < |p_1|$, $p_1, p_2 \in \mathbb{K}$, $\varepsilon > 0$, *and* $\varphi_s : \mathbb{R} \to X$ *satisfy the inequality*

$$\sup_{x \in \mathbb{R}} \|\varphi_s(x + 2) + a_1\varphi_s(x + 1) + a_0\varphi_s(x)\| \le \varepsilon. \tag{5.153}$$

Then there is a unique solution $\varphi : \mathbb{R} \to X$ of the functional equation

$$\varphi(x + 2) + a_1\varphi(x + 1) + a_0\varphi(x) = 0 \tag{5.154}$$

with

$$\sup_{x \in \mathbb{R}} \|\varphi_s(x) - \varphi(x)\| \leq \frac{(|p_1| - |p_2|)\varepsilon}{|p_1 - p_2|(|p_1| - 1)(1 - |p_2|)}. \tag{5.155}$$

Clearly, (5.155) is not worse than (5.150), because

$$|p_1| - |p_2| \leq |p_1 - p_2|;$$

and it is sharper when $|p_1| - |p_2| < |p_1 - p_2|$.

Remark 26. Note that, in the case where a_1, a_0, p_1, p_2 are real numbers, we have $(|p_1| - 1)(|p_2| - 1) = |a_0| + 1 - r_0$ with

$$r_0 := |p_1| + |p_2| = \begin{cases} |a_1| & \text{if } a_0 > 0; \\ \sqrt{a_1^2 - 4a_0} & \text{if } a_0 < 0. \end{cases}$$

8. Linear differential equations

In this part we present a result on stability of ordinary linear differential equations. We consider only the case

$$p_i \in \mathbb{K}, \qquad i = 1, \ldots, m. \tag{5.156}$$

Let I be a nonempty real open interval, $C(I, X)$ be the space of all continuous mappings $\varphi : I \to X$ and $C^n(I, X)$ the subspace of all functions $\varphi : I \to X$ that are n times strongly differentiable (see [25, p. 996]).

In what follows $d(I)$ denotes the length of the interval I. We assume that
(*) $\Re p_i \neq 0$ for $i = 1, \ldots, m$ or $d(I) < \infty$,
where $\Re z$ denotes the real part of the complex number z for $\mathbb{K} = \mathbb{C}$ and $\Re z = z$ for $\mathbb{K} = \mathbb{R}$. According to the results in [25, Remark 1 and Corollaries 2–4] (cf. [31, Theorem 2.1]) and Proposition 4 we can state the following (see [33, Remark 4.1]).

Proposition 10. *Let X be complete, (*) be valid, $i \in \{1, \ldots, m\}$, $\varphi_s, \eta \in C^1(I, X)$, $\delta \in \mathbb{R}_+$ and*

$$\|\varphi_s' - p_i\varphi_s - \eta\|^* \leq \delta.$$

Then there is $\varphi \in C^1(I, X)$ such that

$$\varphi' - p_i\varphi = \eta$$

and

$$\|\varphi_s - \varphi\|^* \le \tau_i \delta,$$

where

$$\tau_i = \begin{cases} d(I), & \text{if } \Re p_i = 0 \text{ and } d(I) < \infty; \\ 1/|\Re p_i|, & \text{if } \Re p_i \neq 0. \end{cases} \tag{5.157}$$

In view of Theorem 10, for each $i \in \{1, \ldots, m\}$, hypothesis $(\mathcal{H}_i)$ holds with $\mathcal{L} = d/dt$, $\mathcal{U} = C^1(I, X)$ and $\rho_i(\delta) = \tau_i \delta$ for $\delta \in \mathbb{R}_+$. Hence from Theorem 89 we deduce (cf. [33, Corollary 4.2]) the following Hyers-Ulam stability result for the equation

$$\varphi^{(m)} + a_{m-1}\varphi^{(m-1)} + \ldots + a_1\varphi' + a_0\varphi = F. \tag{5.158}$$

It has been already proved in [26], but with a different and more involved method (and in a bit weaker form).

Theorem 104. *Suppose that X is complete, condition (5.156) holds, $\delta \in \mathbb{R}_+$, $F \in C(I, X)$, and $\varphi_s \in C^m(I, X)$ satisfies*

$$\left\|\varphi_s^{(m)} + a_{m-1}\varphi_s^{(m-1)} + \ldots + a_1\varphi_s' + a_0\varphi_s - F\right\|^* \le \delta. \tag{5.159}$$

If hypothesis $(*)$ is valid, then there exists a solution $\varphi \in C^m(I, X)$ of equation (5.158) such that

$$\|\varphi_s - \varphi\| \le \delta \prod_{i=1}^{m} \tau_i, \tag{5.160}$$

where τ_i is given by (5.157).

Proof. It is enough to notice that hypothesis $(\mathcal{G})$ holds, because equation (5.158) admits a solution for each $F \in C(I, X)$. $\square$

Similar results for integral equations can be derived form the outcomes in [1, 3, 13, 14, 15, 18, 19, 27, 30]. One example of them is provided in the next section.

9. Integral equations

We end this chapter with a stability result for some integral equation with a delay, patterned on results in [3].

Let t_0 be a fixed real number and I be a real interval of the form $[t_0, a]$, with some $a > t_0$. Let $C_J(\mathbb{K})$ denote the family of all continuous functions mapping a real interval

J into $\mathbb{K}$, $\alpha \in (0, 1)$, $h > 0$ be a real number,

$$I_h := I \cup [t_0 - h, t_0],$$

and $C_{[-h,0]}(\mathbb{K})$ be endowed with the supremum norm denoted by $\| \cdot \|$. Given $y \in C_{I_h}(\mathbb{K})$, we define $y_t \in C_{[-h,0]}(\mathbb{K})$ for $t \in I$ by

$$y_t(\theta) := y(t + \theta), \qquad \theta \in [-h, 0].$$

Next, $b : I \times \mathbb{K} \to \mathbb{K}$ is continuous with respect to the first variable,

$$b(t_0, s) = s, \qquad s \in \mathbb{K}, \tag{5.161}$$

$$I_2 := \{(t, s) \in I^2 : s < t\},$$

$N : I_2 \times C_{[-h,0]}(\mathbb{K}) \to \mathbb{K}$ is continuous, and $L : I_2 \times [0, \infty) \to [0, \infty)$ is nondecreasing with respect to the third variable and continuous.

Moreover, we assume that there is $K \in (0, \infty)$ with

$$\int_{t_0}^t L(t, s, \gamma)ds \le K\gamma, \qquad t \in I, \gamma \in \mathbb{R}_+, \tag{5.162}$$

and

$$|N(t, s, z_s) - N(t, s, w_s)| \le L(t, s, \|z_s - w_s\|), \tag{5.163}$$
$$(t, s) \in I_2, z, w \in C_{I_h}(\mathbb{K}).$$

The following theorem can be easily deduced from [3, Theorem 2.1].

Theorem 105. *Let $K < 1$, $y \in C_{I_h}(\mathbb{K})$, and*

$$\left| y(t) + b(t, y(t_0)) + \int_{t_0}^t N(t, s, y_s)ds \right| \le \delta, \qquad t \in I. \tag{5.164}$$

Then there is a unique solution $\widehat{y} \in C_{I_h}(\mathbb{K})$ of the equation

$$\widehat{y}(t) + b(t, \widehat{y}(t_0)) + \int_{t_0}^t N(t, s, \widehat{y}_s)ds = 0, \qquad t \in I, \tag{5.165}$$

such that

$$|y(t) - \widehat{y}(t)| \le \frac{\delta}{1 - K}, \qquad t \in I, \tag{5.166}$$

$$\widehat{y}(t) = y(t), \qquad t \in [t_0 - h, t_0]. \tag{5.167}$$

Motivated by it we present the subsequent stability result that can be derived form

Theorem 91, for the integral equation

$$(\mathfrak{I}^m \phi)(t) + \sum_{j=0}^{m-1} a_j (\mathfrak{I}^j \phi)(t) = F(t), \qquad (5.168)$$

with

$$(\mathfrak{I}w)(t) = b(t, w(t_0)) + \int_{t_0}^t N(t, s, w_s) ds, \qquad t \in I,$$

$$(\mathfrak{I}w)(t) = w(t), \qquad t \in [t_0 - h, t_0],$$

for every $w \in C_{I_h}(\mathbb{K})$. It corresponds, in particular, to the outcomes in [1, 13, 14, 15, 18, 19, 27, 30].

In what follows we assume additionally that N is linear with respect to the third variable (i.e., the function $N(s, t, \cdot)$ is linear for every $(s, t) \in I_2$) and b is linear with respect to the second variable.

Theorem 106. *Let $F, \varphi_s \in C_{I_h}(\mathbb{K})$, $F(t_0) = 0 = \varphi_s(0)$, and $\delta \in \mathbb{R}_+$ be such that*

$$|P_m(\mathfrak{I})\varphi_s(t) - F(t)| \leq \delta, \qquad t \in I_h. \qquad (5.169)$$

Assume that $|p_i| > \kappa$ for $i = 1, \ldots, m$, where

$$\kappa := \begin{cases} \max\{1, K\}, & \text{if (5.124) holds;} \\ \sqrt{2} \max\{1, K\}, & \text{otherwise.} \end{cases}$$

Then there is a unique solution $\varphi \in C_{I_h}(\mathbb{K})$ of equation (5.168) with $\varphi(t_0) = 0$ and

$$|\varphi_s(t) - \varphi(t)| \leq \frac{\delta}{(|p_1| - \kappa) \ldots (|p_m| - \kappa)}, \qquad t \in I_h. \qquad (5.170)$$

Proof. Let $\mathcal{U} := \{w \in C_{I_h}(\mathbb{K}) : w(t_0) = 0\}$. Clearly, by (5.161), $\mathfrak{I}(\mathcal{U}) \subset \mathcal{U}$. Next, take $z, w \in \mathcal{U}$. Then, from (5.162) and (5.163), we obtain that

$$|(\mathfrak{I}z)(t) - (\mathfrak{I}w)(t)| \leq \int_{t_0}^t L(t, s, \|z_s - w_s\|) ds$$

$$\leq \int_{t_0}^t L(t, s, \|z - w\|_\infty) ds \leq K \|z - w\|_\infty, \qquad t \in I,$$

where $\| \cdot \|_\infty$ stands for the supremum norm in $C_{I_h}(\mathbb{K})$. Moreover,

$$|(\mathfrak{I}z)(t) - (\mathfrak{I}w)(t)| = |z(t) - w(t)|, \qquad t \in [t_0 - h, t_0].$$

Thus we have obtained that $\|\mathfrak{I}\| \leq K_0 := \max\{1, K\}$. Clearly, by Proposition 5,

$$\|\widehat{\mathfrak{I}}\|_T \leq \sqrt{2} K_0.$$

Consequently, Theorem 91 completes the proof.   □

REFERENCES

1. A. Bahyrycz, J. Brzdęk, Z. Leśniak, On approximate solutions of the generalized Volterra integral equation, Nonlinear Anal. RWA 20 (2014) 59–66.
2. J. Brzdęk, Remarks on stability of some inhomogeneous functional equations, Aequationes Math. 89 (2015) 83–96.
3. J. Brzdęk, N. Eghbali, On approximate solutions of some delayed fractional differential equations, Appl. Math. Lett. 54 (2016) 31–35.
4. J. Brzdęk, S.M. Jung, A note on stability of a linear functional equation of second order connected with the Fibonacci numbers and Lucas sequences, J. Inequal. Appl. 2010 (2010), Art. ID 793947, 10 pp.
5. J. Brzdęk, S.M. Jung, A note on stability of an operator linear equation of the second order, Abstr. Appl. Anal. 2011 (2011), Art. ID 602713, 15 pp.
6. J. Brzdęk, D. Popa, B. Xu, Note on the nonstability of the linear recurrence, Abh. Math. Sem. Univ. Hamburg 76 (2006) 183–189.
7. J. Brzdęk, D. Popa, B. Xu, The Hyers-Ulam stability of nonlinear recurrences, J. Math. Anal. Appl. 335 (2007) 443–449.
8. J. Brzdęk, D. Popa, B. Xu, Hyers-Ulam stability of linear equations of higher orders, Acta Math. Hungar. 120 (2008) 1–8.
9. J. Brzdęk, D. Popa, B. Xu, Remarks on stability of linear recurrence of higher order, Appl. Math. Lett. 23 (2010) 1459–1463.
10. J. Brzdęk, D. Popa, B. Xu, On nonstability of the linear recurrence of order one, J. Math. Anal. Appl. 367 (2010) 146–153.
11. J. Brzdęk, D. Popa, B. Xu, Note on nonstability of the linear functional equation of higher order, Comput. Math. Appl. 62 (2011) 2648–2657.
12. J. Brzdęk, D. Popa, B. Xu, On approximate solutions of the linear functional equation of higher order, J. Math. Anal. Appl. 373 (2011) 680–689.
13. L. Cădariu, V. Radu, Fixed point methods for the generalized stability of functional equations in a single variable, Fixed Point Theory Appl. 2008 (2008) Art. ID 749392, 15 pp.
14. L.P. Castro, A. Ramos, Hyers-Ulam-Rassias stability for a class of nonlinear Volterra integral equations, Banach J. Math. Anal. 3 (2009) 36–43.
15. L.P. Castro, A. Ramos, Hyers-Ulam and Hyers-Ulam-Rassias stability of Volterra integral equations with delay, In: Integral Methods in Science and Engineering, Vol. 1, pp. 85–94, Birkhäuser, Boston, 2010.
16. J.B. Diaz, B. Margolis, A fixed point theorem of the alternative, for contractions on a generalized complete metric space, Bull. Amer. Math. Soc. 74 (1968) 305–309.
17. M. Fabian, P. Habala, P. Hájek, V. Montesinos Santalucia, J. Pelant, V. Zizler, Functional Analysis and Infinite-Dimensional Geometry, Springer-Verlag, New York, 2001.
18. M. Gachpazan, O. Baghani, Hyers–Ulam stability of nonlinear integral equation. Fixed Point Theory Appl. 2010 (2010), Art. ID 927640, 6 pp.
19. S.M. Jung, A fixed point approach to the stability of a Volterra integral equation. Fixed Point Theory Appl. 2007 (2007), Art. ID 57064, 9 pp.
20. S.M. Jung, Functional equation $f(x) = pf(x - 1) - qf(x - 2)$ and its Hyers-Ulam stability, J. Inequal. Appl. 2009 (2009), Art. ID 181678, 10 pp.
21. S.M. Jung, Hyers-Ulam stability of Fibonacci functional equation, Bull. Iran. Math. Soc. 35 (2009) 217–227.
22. R.V. Kadison, J.R. Ringrose, Fundamentals of the Theory of Operator Algebras, v. I: Elementary Theory. Reprint of the 1983 original. American Mathematical Society, Providence, RI, 1997.
23. M. Kuczma, Functional Equations in a Single Variable, Polish Scientific Publishers, Warszawa, 1968.
24. M. Kuczma, B. Choczewski, R. Ger, Iterative Functional Equations, Encyclopedia of Mathematics and its Applications, Cambridge University Press, 1990.

25. T. Miura, S.M. Jung, S.E. Takahasi, Hyers-Ulam-Rassias stability of the Banach space valued linear differential equations $y' = \lambda y$, J. Korean Math. Soc. 41 (2004) 995–1005.
26. T. Miura, S. Miyajima, S.E. Takahasi, Hyers-Ulam stability of linear differential operator with constant coefficients, Math. Nachr. 258 (2003) 90–96.
27. J.R. Morales, E.M. Rojas, Hyers-Ulam and Hyers-Ulam-Rassias stability of nonlinear integral equations with delay. Int. J. Nonlinear Anal. Appl. 2 (2011) 1–6.
28. D. Popa, Hyers-Ulam stability of the linear recurrence with constant coefficients, Adv. Difference Equ. 2005 (2005) 101–107.
29. D. Popa, Hyers-Ulam-Rassias stability of a linear recurrence, J. Math. Anal. Appl. 309 (2005) 591–597.
30. I.A. Rus, Gronwall lemma approach to the Hyers-Ulam-Rassias stability of an integral equation. In: Nonlinear Analysis and Variational Problems. Springer Optim. Appl. v. 35, pp. 147–152, Springer, New York, 2010.
31. S.E. Takahasi, T. Miura, S. Miyajima, On the Hyers-Ulam stability of the Banach space-valued differential equation $y' = \lambda y$, Bull. Korean Math. Soc. 39 (2002) 309–315.
32. B. Xu, J. Brzdęk, Hyers-Ulam stability of a system of first order linear recurrences with constant coefficients, Discrete Dyn. Nat. Soc. 2015, Art. ID 269356, 5 pp.
33. B. Xu, J. Brzdęk, W. Zhang, Fixed-point results and the Hyers-Ulam stability of linear equations of higher orders, Pacific J. Math 273 (2015) 483–498.

Contents

Abstract

In this chapter we deal with some possible approaches to study the situations where there is a lack of stability in some sense (we frequently say that nonstability occurs). Several quite general results are already known (concerning mainly the stability of difference and functional equations); however, in general the subject of nonstability has not yet been studied extensively. One of the main reasons for this situation is that it is not easy to formulate in a precise form a general definition of the notion of nonstability, because not only must it depend on the control function that one might use, but also on what we expect to obtain in the final estimate. The situation is somewhat simpler for the case of Hyers-Ulam stability, which occurs when the control function is constant (one expects the same phenomenon in the final setting).

1. Preliminary information

In this chapter, analogously as in the previous one, $\mathbb{K}$ is either the field $\mathbb{R}$ or $\mathbb{C}$, X is a normed space over $\mathbb{K}$, and $\mathbb{S} := \{a \in \mathbb{C} : |a| = 1\}$, unless explicitly stated otherwise. Moreover, $\mathbb{Z}$ and $\mathbb{N}$ denote the sets of integers and positive integers, respectively; $\mathbb{N}_0 := \mathbb{N} \cup \{0\}$ and $\mathbb{R}_+ := [0, \infty)$.

We deal with some possible approaches to investigate the situations where there is a lack of stability in certain sense; we simply say that nonstability occurs (in that sense). Several quite general results on nonstability are already known (concerning mainly the stability of difference and functional equations); however, in general the subject of nonstability has not yet been studied extensively. One of the main reasons for this situation is that it is not easy to formulate precisely a general definition of the notion of nonstability, because it must somehow depend on the control function that is applied and on what we expect to obtain as the final estimate.

The first known result, that we could call a nonstability theorem and which shows what we might understand by nonstability of an equation, has been proved in [9] and concerns the Cauchy functional equation:

$$f(x + y) = f(x) + f(y). \tag{6.1}$$

It is connected with Theorem 81 (in Chapter 4) and states (see [9]; cf. also [11]) that for $p = 1$ an analogous result is not valid. Namely, we have the subsequent:

Theorem 107. *Let $c \in \mathbb{R}$, $c > 0$. There is $f : \mathbb{R} \to \mathbb{R}$ such that*

$$|f(x + y) - f(x) - f(y)| \leq c(|x| + |y|), \qquad x, y \in \mathbb{R}, \tag{6.2}$$

and

$$\sup_{x \in \mathbb{R} \setminus \{0\}} \frac{|f(x) - h(x)|}{|x|} = \infty \tag{6.3}$$

for each solution $h : \mathbb{R} \to \mathbb{R}$ of (6.1).

It can be easily generalized for the inhomogeneous version of equation (6.1), i.e., for the equation

$$g(x + y) = g(x) + g(y) + d(x, y), \tag{6.4}$$

with a given function $d : \mathbb{R}^2 \to \mathbb{R}$, in the following way (cf. [2]).

Theorem 108. *Let $d : \mathbb{R}^2 \to \mathbb{R}$. Assume that (6.4) admits a solution $f_0 : \mathbb{R} \to \mathbb{R}$. Then for each real $c_0 > 0$ there is $f : \mathbb{R} \to \mathbb{R}$ such that*

$$|f(x + y) - f(x) - f(y) - d(x, y)| \leq c_0(|x| + |y|), \qquad x, y \in \mathbb{R}, \tag{6.5}$$

and

$$\sup_{x \in \mathbb{R} \setminus \{0\}} \frac{|f(x) - g(x)|}{|x|} = \infty \tag{6.6}$$

for each solution $g : \mathbb{R} \to \mathbb{R}$ of (6.4).

Proof. Fix $c_0 > 0$. According to Theorem 107, there is a function $f_1 : \mathbb{R} \to \mathbb{R}$ with

$$|f_1(x + y) - f_1(x) - f_1(y)| \leq |x| + |y|, \qquad x, y \in \mathbb{R},$$

and such that, for each additive $h_0 : \mathbb{R} \to \mathbb{R}$,

$$\sup_{x \in \mathbb{R} \setminus \{0\}} \frac{|f_1(x) - h_0(x)|}{|x|} = \infty.$$

Let $f := c_0 f_1 + f_0$. Clearly

$$|f(x + y) - f(x) - f(y) - d(x, y)| = c_0 |f_1(x + y) - f_1(x) - f_1(y)|$$
$$\leq c_0(|x| + |y|), \qquad x, y \in \mathbb{R}.$$

Next, fix a solution $h : \mathbb{R} \to \mathbb{R}$ of (6.4). Then $h_1 := h - f_0$ is additive and

$$\sup_{x \in \mathbb{R} \setminus \{0\}} \frac{|f(x) - h(x)|}{|x|} = \sup_{x \in \mathbb{R} \setminus \{0\}} c_0 \frac{|f_1(x) - c_0^{-1} h_1(x)|}{|x|} = \infty.$$

$\square$

Remark 27. The assumption of Theorem 108 that equation (6.4) admits a solution $f_0 : \mathbb{R} \to \mathbb{R}$ is quite natural, because it seems to be a bit strange to investigate stability of an equation of which we know that it does not have any solution. The next two examples show that this may happen quite naturally and that the issue is not trivial.

Example 15. Let $d : \mathbb{R}^2 \to \mathbb{R}$, $c, p \in \mathbb{R}$, $p < 0$, $c > 0$, $d(x_0, y_0) \neq 0$ for some $x_0, y_0 \in \mathbb{R}$, and

$$d(x, y) \leq c(|x|^p + |y|^p), \qquad x, y \in \mathbb{R} \setminus \{0\}.$$

Then, functional equation (6.4) has no solutions $g : \mathbb{R} \to \mathbb{R}$. In fact, suppose that $g : \mathbb{R} \to \mathbb{R}$ is a solution to (6.4). Then

$$|g(x + y) - g(x) - g(y)| \leq c(|x|^p + |y|^p), \qquad x, y \in \mathbb{R} \setminus \{0\},$$

whence Theorem 81 implies that g must be additive and consequently

$$d(x, y) = g(x + y) - g(x) - g(y) = 0, \qquad x, y \in \mathbb{R}.$$

This is a contradiction to the assumption that $d(x_0, y_0) \neq 0$.

But, on the other hand, for every additive function $h : \mathbb{R} \to \mathbb{R}$, we have

$$|h(x + y) - h(x) - h(y) - d(x, y)| = |d(x, y)|$$
$$\leq c(|x|^p + |y|^p), \qquad x, y \in \mathbb{R} \setminus \{0\}.$$

Example 16. Let $c \in \mathbb{R}$, $c > 0$, and $d : \mathbb{R}^2 \to \mathbb{R}$ be given by the following:

$$d(x, y) = c(|x| + |y|), \qquad x, y \in \mathbb{R}.$$

Then, for any additive function $h : \mathbb{R} \to \mathbb{R}$, we have

$$|h(x+y) - h(x) - h(y) - d(x,y)| = |d(x,y)| = c(|x| + |y|), \qquad x, y \in \mathbb{R}.$$

Suppose that equation (6.4) has a solution $g : \mathbb{R} \to \mathbb{R}$. Then

$$d(x,y) = g(x+y) - g(x) - g(y), \qquad x, y \in \mathbb{R},$$

whence

$$
\begin{aligned}
d(x+y, z) + d(x, y) &= g(x+y+z) - g(x+y) - g(z) \\
&\quad + g(x+y) - g(x) - g(y) \\
&= g(x+y+z) - g(z) - g(x) - g(y) \\
&= g(x+y+z) - g(x) - g(y+z) \\
&\quad + g(y+z) - g(y) - g(z) \\
&= d(x, y+z) + d(y, z), \qquad x, y, z \in \mathbb{R}.
\end{aligned}
$$

It is easy to check that this yields that

$$|x + y| = |y + z|, \qquad x, y, z \in \mathbb{R},$$

which is not true. Thus we have proved that for such d, (6.4) does not have any solution $g : \mathbb{R} \to \mathbb{R}$.

2. Possible definitions of nonstability

It is somewhat easier to define the notion of nonstability in the case of Hyers-Ulam stability, which occurs when the control function is constant and one expects the same phenomenon in the final estimate. For example, we could state the following definition:

Definition 9. Let P and D be nonempty sets and (Y, d) and (Z, ρ) be metric spaces. Let $\mathcal{V} \subset \mathcal{U} \subset Y^D$ be nonempty, $F \in Z^P$ and $\mathcal{T} : \mathcal{U} \to Z^P$. We say that the equation

$$\mathcal{T}\varphi = F, \tag{6.7}$$

is weakly nonstable in $\mathcal{V}$ in the Hyers-Ulam sense, provided there is a function $f \in \mathcal{V}$ such that

$$\sup_{x \in P} \rho((\mathcal{T} f)(x), F(x)) < \infty \tag{6.8}$$

and, for each solution $\gamma \in \mathcal{U}$ of (6.7), we have

$$\sup_{x \in D} d(f(x), \gamma(x)) = \infty.$$

That definition corresponds to the next definition stated in [3] and concerning the linear difference equations in X with constant coefficients of the form

$$x_{n+m} + a_{m-1}x_{n+m-1} + \ldots + a_0 x_n = b_n, \qquad n \in \mathbb{N}_0, \tag{6.9}$$

where $m \in \mathbb{N}$ and $a_0, \ldots, a_{m-1} \in \mathbb{K}$ are fixed and $(b_n)_{n\in\mathbb{N}_0}$ is a given sequence in X.

Definition 10. Difference equation (6.9) is said to be Hyers-Ulam weakly stable provided, for every unbounded sequence $(y_n)_{n\in\mathbb{N}_0}$ in X with

$$\sup_{n\in\mathbb{N}_0} \|y_{n+m} + a_{m-1}y_{n+m-1} + \ldots + a_0 y_n - b_n\| < \infty, \tag{6.10}$$

there exists a sequence $(x_n)_{n\in\mathbb{N}_0}$ in X such that (6.9) holds and

$$\sup_{n\in\mathbb{N}_0} \|y_n - x_n\| < \infty.$$

But also the following definition has been formulated in [3]; it is consistent with the original formulation of the Ulam stability problem in [10].

Definition 11. Difference equation (6.9) is said to be Hyers-Ulam strongly stable provided, for every $\varepsilon > 0$, there exists $\delta > 0$ such that, for every sequence $(y_n)_{n\in\mathbb{N}_0}$ in X satisfying

$$\sup_{n\in\mathbb{N}_0} \|y_{n+m} + a_{m-1}y_{n+m-1} + \ldots + a_0 y_n - b_n\| < \delta, \tag{6.11}$$

there exists a sequence $(x_n)_{n\in\mathbb{N}_0}$ in X such that (6.9) holds and

$$\sup_{n\in\mathbb{N}_0} \|y_n - x_n\| \le \varepsilon.$$

Clearly, if a difference equation is strongly stable, then it is also weakly stable. The next definition is even somewhat stronger than Definition 11, but it corresponds to the notion of Hyers-Ulam stability already used in previous chapters.

Definition 12. Let $\kappa \in \mathbb{R}_+$. Difference equation (6.9) is said to be Ulam κ−stable provided, for every sequence $(\delta_n)_{n\in\mathbb{N}_0}$ in $\mathbb{R}_+$ and every sequence $(y_n)_{n\in\mathbb{N}_0}$ in X satisfying

$$\|y_{n+m} + a_{m-1}y_{n+m-1} + \ldots + a_0 y_n - b_n\| < \delta_n, \qquad n \in \mathbb{N}_0, \tag{6.12}$$

there exists a sequence $(x_n)_{n\in\mathbb{N}_0}$ in X such that (6.9) holds and

$$\|y_{n+1} - x_{n+1}\| \le \kappa\delta_n, \qquad n \in \mathbb{N}_0.$$

The nonstability results contained in the next section will refer mainly to the notion of stability described in Definition 12.

3. Linear difference equation of the first order

Let $(b_n)_{n\in\mathbb{N}_0}$ be a sequence in X, $(a_n)_{n\in\mathbb{N}_0}$ be a sequence in $\mathbb{K}\setminus\{0\}$, and $(\varepsilon_n)_{n\in\mathbb{N}_0}$ be a sequence of positive real numbers. It is already known from [19] (see also [4, 5, 20]) that, in the case where

$$\limsup_{n\to\infty} \frac{\varepsilon_n|a_{n+1}|}{\varepsilon_{n+1}} < 1 \quad \text{or} \quad \liminf_{n\to\infty} \frac{\varepsilon_n|a_{n+1}|}{\varepsilon_{n+1}} > 1, \tag{6.13}$$

for every sequence $(y_n)_{n\in\mathbb{N}_0}$ in X satisfying the relation

$$\|y_{n+1} - a_n y_n - b_n\| \le \varepsilon_n, \qquad n \in \mathbb{N}_0, \tag{6.14}$$

there exists a sequence $(x_n)_{n\in\mathbb{N}_0}$ in X such that

$$x_{n+1} = a_n x_n + b_n, \qquad n \in \mathbb{N}_0, \tag{6.15}$$

and

$$L := \sup_{n\in\mathbb{N}} \frac{\|y_n - x_n\|}{\varepsilon_{n-1}} < \infty, \tag{6.16}$$

which means that

$$\|y_n - x_n\| \le L\varepsilon_{n-1}, \qquad n \in \mathbb{N}.$$

In connection with this property there arises a natural question whether condition (6.13) can be weakened. Simple examples given in [19] show that if (6.13) does not hold, then analogous result is not generally true.

Following the terminology in the previous chapters we introduce the subsequent definition:

Definition 13. We say that difference equation (6.15) is Ulam $(\varepsilon_n)_{n\in\mathbb{N}_0}$-weakly stable provided, for every sequence $(y_n)_{n\in\mathbb{N}_0}$ in X satisfying (6.14), there exists a sequence $(x_n)_{n\in\mathbb{N}_0}$ in X such that (6.15) and (6.16) hold. Otherwise, we say that the difference equation is Ulam $(\varepsilon_n)_{n\in\mathbb{N}_0}$-weakly nonstable.

In this section we investigate stability of (6.15) in the case where condition (6.13) can be possibly not valid. First we consider the situation when

$$\lim_{n\to\infty} \frac{\varepsilon_n|a_{n+1}|}{\varepsilon_{n+1}} = 1. \tag{6.17}$$

We show that then difference equation (6.15) is Ulam $(\varepsilon_n)_{n\in\mathbb{N}_0}$-weakly nonstable (see Corollary 12).

Let us start with the following simple observation.

Lemma 10. *Let $(z_n)_{n \geq 0}$ and $(d_n)_{n \in \mathbb{N}_0}$ be sequences in X and*

$$z_{n+1} = a_n z_n + d_n, \qquad n \in \mathbb{N}_0.$$

Then

$$z_n = a_0 \ldots a_{n-1} z_0 + \sum_{k=1}^{n-1} a_k \ldots a_{n-1} d_{k-1} + d_{n-1}, \qquad n \geq 2. \tag{6.18}$$

Moreover, if $(y_n)_{n \in \mathbb{N}_0}$ is a sequence in X with

$$\|y_{n+1} - a_n y_n - d_n\| \leq \varepsilon_n, \qquad n \in \mathbb{N}_0,$$

then

$$
\begin{aligned}
\|y_n - z_n\| \quad \leq \quad & |a_0 \ldots a_{n-1}| \, \|y_0 - z_0\| \\
& + \sum_{k=1}^{n-1} |a_k \ldots a_{n-1}| \varepsilon_{k-1} + \varepsilon_{n-1}, \qquad n \geq 2.
\end{aligned}
\tag{6.19}
$$

Proof. By induction on n it is easy to show that (6.18) holds. Next, let $(y_n)_{n \in \mathbb{N}_0}$ be a sequence in X such that

$$\|y_{n+1} - a_n y_n - d_n\| \leq \varepsilon_n, \qquad n \in \mathbb{N}_0.$$

Write

$$c_n := y_{n+1} - a_n y_n - d_n, \qquad n \in \mathbb{N}_0. \tag{6.20}$$

Then

$$y_{n+1} = a_n y_n + d_n + c_n, \qquad n \in \mathbb{N}_0,$$

and, by (6.18) with z_n and d_n replaced by y_n and $d_n + c_n$, respectively, we obtain

$$y_n = a_0 \ldots a_{n-1} y_0 + \sum_{k=1}^{n-1} a_k \ldots a_{n-1} (d_{k-1} + c_{k-1}) + d_{n-1} + c_{n-1}$$

for $n \geq 2$, which implies (6.19). $\qquad \square$

Fix $y, y_0 \in X$, $0 < \|y\| \leq 1$, and write

$$c_n := \varepsilon_n \frac{a_0 \ldots a_n}{|a_0 \ldots a_n|} y, \qquad n \in \mathbb{N}_0. \tag{6.21}$$

Let $(y_n)_{n \in \mathbb{N}_0}$ be a sequence in X given by

$$y_{n+1} = a_n y_n + b_n + c_n, \quad n \in \mathbb{N}_0. \tag{6.22}$$

Then it is easily seen that (6.14) holds. The next theorem deals with the situations when, for a given sequence $(x_n)_{n \in \mathbb{N}_0}$ in X satisfying (6.15), we have

$$\sup_{n \in \mathbb{N}} \frac{\|y_n - x_n\|}{\varepsilon_{n-1}} = \infty. \tag{6.23}$$

Theorem 109. *Let $(x_n)_{n \in \mathbb{N}_0}$ be a sequence in X satisfying (6.15) and*

$$\sigma_0 = \limsup_{n \to \infty} \frac{|a_0 \ldots a_{n-1}|}{\varepsilon_{n-1}}, \qquad s := \lim_{n \to \infty} \sum_{k=1}^{n} \frac{\varepsilon_{k-1}}{|a_0 \ldots a_{k-1}|},$$

Assume that one of the following four conditions is fulfilled:

(a) *$s < \infty$ and $x_0 \neq y_0 + sy$;*
(b) *$s < \infty$, $x_0 = y_0 + sy$ and (6.17) is valid;*
(c) *$s = \infty$ and $\sigma_0 \neq 0$;*
(d) *$s = \infty$, $\sigma_0 = 0$ and (6.17) is valid.*

Then (6.23) holds.

Proof. Write

$$s_n := \sum_{k=1}^{n} \frac{\varepsilon_{k-1}}{|a_0 \ldots a_{k-1}|}, \qquad n \geq 1. \tag{6.24}$$

Clearly $s := \lim_{n \to \infty} s_n$. On account of Lemma 10 with $z_n := y_n - x_n$ and $d_n := c_n$, for each $n \in \mathbb{N}$ we get

$$y_n - x_n = a_0 \ldots a_{n-1} \left(y_0 - x_0 + \sum_{k=1}^{n} \frac{c_{k-1}}{a_0 \ldots a_{k-1}} \right)$$

$$= a_0 \ldots a_{n-1} (y_0 - x_0 + s_n y). \tag{6.25}$$

First assume that $s < \infty$. Then

$$\lim_{n \to \infty} \frac{\varepsilon_{n-1}}{|a_0 \ldots a_{n-1}|} = 0,$$

whence

$$\lim_{n \to \infty} \frac{|a_0 \ldots a_{n-1}|}{\varepsilon_{n-1}} = \infty. \tag{6.26}$$

Further,

$$\lim_{n \to \infty} \left\| y_0 - x_0 + \sum_{k=1}^{n} \frac{c_{k-1}}{a_0 \ldots a_{k-1}} \right\| = \lim_{n \to \infty} \|y_0 - x_0 + s_n y\|$$

$$= \|y_0 - x_0 + sy\|$$

and consequently

$$\lim_{n\to\infty} \frac{\|y_n - x_n\|}{\varepsilon_{n-1}} = \lim_{n\to\infty} \frac{|a_0 \ldots a_{n-1}|}{\varepsilon_{n-1}} \|y_0 - x_0 + sy\|. \qquad (6.27)$$

This ends the proof when $x_0 \neq y_0 + sy$, because (6.26) and (6.27) imply that

$$\lim_{n\to\infty} \frac{\|y_n - x_n\|}{\varepsilon_{n-1}} = \infty.$$

So, let now $x_0 = y_0 + sy$. Then, for each $n \in \mathbb{N}$,

$$\begin{aligned}
\frac{\|y_n - x_n\|}{\varepsilon_{n-1}} &= \frac{|a_0 \ldots a_{n-1}|}{\varepsilon_{n-1}} \|(s - s_n)y\| \\
&= \frac{|a_0 \ldots a_{n-1}|}{\varepsilon_{n-1}} (s - s_n)\|y\| = \frac{s - s_n}{s_n - s_{n-1}} \|y\| \\
&= \frac{\|y\|}{\sigma_n - 1},
\end{aligned}$$

where

$$\sigma_n := \frac{s - s_{n-1}}{s - s_n}, \qquad n \in \mathbb{N}.$$

Next, according to the well-known Stolz-Cesaro lemma (the case 0/0),

$$\begin{aligned}
\lim_{n\to\infty} \frac{s - s_{n-1}}{s - s_n} &= \lim_{n\to\infty} \frac{(s - s_n) - (s - s_{n-1})}{(s - s_{n+1}) - (s - s_n)} \\
&= \lim_{n\to\infty} \frac{s_{n-1} - s_n}{s_n - s_{n+1}} = \lim_{n\to\infty} \frac{\varepsilon_{n-1}|a_n|}{\varepsilon_n},
\end{aligned}$$

whence, in the case where (6.17) is valid, we have

$$\lim_{n\to\infty} \sigma_n = 1,$$

which yields (6.23). Thus we have considered the cases of (a) and (b).

Now, suppose that $s = \infty$. If $\sigma_0 \neq 0$, then (6.23) is fulfilled (in view of (6.25)), because

$$\lim_{n\to\infty} \|y_0 - x_0 + s_n y\| = \infty.$$

This completes the proof in the situation when (c) holds.

So, it remains to consider the case when $\sigma_0 = 0$ and (6.17) is valid. The equality $\sigma_0 = 0$ implies that

$$\lim_{n\to\infty} \alpha_n = 0,$$

where

$$\alpha_n := \frac{a_0 \dots a_n}{\varepsilon_n}, \qquad n \in \mathbb{N}_0.$$

Clearly, from (6.25) it follows that, for every $n \in \mathbb{N}$,

$$\frac{y_n - x_n}{\varepsilon_{n-1}} = \alpha_{n-1}(y_0 - x_0) + \alpha_{n-1} s_n y. \tag{6.28}$$

Write

$$z_n := \alpha_{n-1} s_n y, \qquad n \in \mathbb{N}_0.$$

Then, in view of (6.17) and the monotonicity and unboundedness of $(s_n)_{n \in \mathbb{N}_0}$, by the Stolz-Cesaro lemma (the case ∞/∞) we get

$$\lim_{n \to \infty} \frac{\|y\|}{\|z_n\|} = \lim_{n \to \infty} \frac{1}{|\alpha_{n-1} s_n|} = \lim_{n \to \infty} \frac{1/|\alpha_n| - 1/|\alpha_{n-1}|}{s_{n+1} - s_n}$$

$$= \lim_{n \to \infty} \left(1 - \frac{|a_n|\varepsilon_{n-1}}{\varepsilon_n} \right) = 0.$$

Therefore

$$\lim_{n \to \infty} \|z_n\| = \infty.$$

Consequently, in view of (6.28), this means that

$$\lim_{n \to \infty} \frac{\|y_n - x_n\|}{\varepsilon_{n-1}} = \infty.$$

$\square$

Theorem 109 yields at once the following.

Corollary 12. *Assume that condition (6.17) holds. Then there exists a sequence $(y_n)_{n \in \mathbb{N}_0}$ in X satisfying (6.14) and such that, for every sequence $(x_n)_{n \in \mathbb{N}_0}$ in X given by (6.15), (6.23) holds.*

Remark 28. It is easily seen that, changing the vector y in (6.21), we obtain a different example of a sequence $(y_n)_{n \in \mathbb{N}_0}$ in X satisfying (6.14) and (6.23). This means that the cardinality of the class of all such sequences $(y_n)_{n \in \mathbb{N}_0}$ in X, occurring in Corollary 12, is equal to the cardinality of the space X.

It seems to be very natural to ask if it is possible to replace condition (6.17) in

Corollary 12 or in Theorem 109 (b), (d) by one of the following two equalities:

$$\liminf_{n\to\infty} \frac{|a_{n+1}|\varepsilon_n}{\varepsilon_{n+1}} = 1, \tag{6.29}$$

$$\limsup_{n\to\infty} \frac{|a_{n+1}|\varepsilon_n}{\varepsilon_{n+1}} = 1. \tag{6.30}$$

Probably, in the general situation, there is no simple answer to that question as the next four examples show (cf. [5, Examples 1–4]). The first one shows that (6.17) in (b) cannot be replaced by (6.29).

Example 17. Let ε be a positive real number, $\varepsilon_n = \varepsilon$, $a_{2n} = 1$, and $a_{2n+1} = 2$ for $n \in \mathbb{N}_0$. Then it is easily seen that $s < \infty$ and

$$\liminf_{n\to\infty} \frac{|a_{n+1}|\varepsilon_n}{\varepsilon_{n+1}} = 1.$$

We show that difference equation (6.15) is $(\varepsilon_n)_{n\in\mathbb{N}_0}$-stable.

Let $(y_n)_{n\in\mathbb{N}_0}$ be a sequence in X such that (6.14) holds and

$$c_n := y_{n+1} - a_n y_n - b_n, \qquad n \in \mathbb{N}_0.$$

(For instance, $(y_n)_{n\in\mathbb{N}_0}$ can be given by (6.22), with c_n defined by (6.21).)

Clearly, there exists

$$\widetilde{s} := \sum_{n=1}^{\infty} \frac{c_{n-1}}{a_0 \ldots a_{n-1}}.$$

Let $x_0 := y_0 + \widetilde{s}$ and

$$x_{n+1} = a_n x_n + b_n, \qquad n \in \mathbb{N}_0.$$

Then (cf. (6.25))

$$\|y_n - x_n\| = |a_0 \ldots a_{n-1}| \left\| -\widetilde{s} + \sum_{k=1}^{n} \frac{c_{k-1}}{a_0 \ldots a_{k-1}} \right\|$$

$$= |a_0 \ldots a_{n-1}| \left\| \sum_{k=n}^{\infty} \frac{c_k}{a_0 \ldots a_k} \right\| \leq M_n \varepsilon,$$

where

$$M_n = \sum_{k=0}^{\infty} \frac{1}{|a_n \ldots a_{n+k}|} = \begin{cases} 3, & \text{if } n \text{ is even;} \\ 2, & \text{if } n \text{ is odd.} \end{cases}$$

The next example shows that (6.17) in (b) cannot be replaced by (6.30).

Example 18. Let ε be a positive real number, $\varepsilon_n = \varepsilon$ for $n \in \mathbb{N}_0$,

$$a_{2n} = \frac{1}{2}, \qquad a_{2n+1} = 1, \qquad n \in \mathbb{N}_0.$$

Then (6.30) holds.

Let $(y_n)_{n \in \mathbb{N}_0}$ be an arbitrary sequence satisfying (6.14), $(x_n)_{n \in \mathbb{N}_0}$ be a sequence satisfying (6.15) with $x_0 := y_0$, and

$$c_n := y_{n+1} - a_n y_n - b_n, \qquad n \in \mathbb{N}_0.$$

Then $\|c_n\| \leq \varepsilon$ for $n \in \mathbb{N}_0$ and Lemma 10 implies (6.18) with

$$z_n := y_n - x_n, \qquad d_n := c_n, \qquad n \in \mathbb{N}_0.$$

Consequently

$$\|y_n - x_n\| \leq |a_0 \ldots a_{n-1}| \left\| \sum_{k=1}^{n} \frac{c_{k-1}}{a_0 \ldots a_{k-1}} \right\|$$

$$\leq \varepsilon |a_0 \ldots a_{n-1}| \sum_{k=1}^{n} \frac{1}{|a_0 \ldots a_{k-1}|}, \qquad n \in \mathbb{N}.$$

This means that, for every $n \in \mathbb{N}$,

$$\|y_{2n} - x_{2n}\| \leq \varepsilon \frac{1}{2^n}(2 + 2 + 2^2 + 2^2 + \ldots + 2^n + 2^n)$$

$$= 4\varepsilon \frac{2^n - 1}{2^n} < 4\varepsilon,$$

$$\|y_{2n+1} - x_{2n+1}\| \leq \varepsilon \frac{1}{2^{n+1}}(2 + 2 + 2^2 + 2^2 + \ldots + 2^n + 2^n + 2^{n+1})$$

$$= 4\varepsilon \frac{2^n - 1}{2^{n+1}} + \varepsilon < 3\varepsilon.$$

This means that, in this situation, difference equation (6.15) is $(\varepsilon_n)_{n \in \mathbb{N}_0}$-stable.

The following examples show that in some situations, when (6.30) holds and (6.17) does not, we also can obtain nonstability of (6.15).

Example 19. Let $\varepsilon_n = \varepsilon$ for $n \in \mathbb{N}_0$, with some $\varepsilon > 0$,

$$I := \{2^k - 2 : k \in \mathbb{N}\},$$

$$a_n = \frac{1}{2}, \qquad n \in I, \qquad a_n = 1, \qquad n \notin I.$$

Then (6.30) holds.

Define $(y_n)_{n\in\mathbb{N}_0}$ by (6.22), where $(c_n)_{n\in\mathbb{N}_0}$ is given by (6.21) and $(x_n)_{n\in\mathbb{N}_0}$ is a sequence in X such that (6.15) holds. Then Lemma 10 (with $z_n := y_n - x_n$ and $d_n := c_n$ for $n \in \mathbb{N}_0$) implies that

$$
\begin{aligned}
y_{2^n-2} - x_{2^n-2} &= a_0 \ldots a_{2^n-3}\left(y_0 - x_0 + \varepsilon y \sum_{k=1}^{2^n-2} \frac{1}{|a_0 \ldots a_{k-1}|}\right) \\
&= \frac{1}{2^{n-1}}(y_0 - x_0 + \varepsilon y(2 + 2 + (2^2 + 2^2 + 2^2 + 2^2) \\
&\qquad + \ldots + (\underbrace{2^{n-1} + \ldots + 2^{n-1}}_{2^{n-1}\ \text{terms}}))) \\
&= \frac{1}{2^{n-1}}(y_0 - x_0 + \varepsilon y(2^2 + 2^4 + \ldots + 2^{2n-2})) \\
&= \frac{1}{2^{n-1}}\left(y_0 - x_0 + 2^2 \cdot \frac{2^{2n-2} - 1}{3} \cdot \varepsilon y\right), \qquad n \in \mathbb{N}.
\end{aligned}
$$

Hence,

$$
\lim_{n\to\infty} \|y_{2^n-2} - x_{2^n-2}\| = \infty.
$$

This means that difference equation (6.15) is $(\varepsilon_n)_{n\in\mathbb{N}_0}$-nonstable.

The last example presents a situations when (c) is fulfilled.

Example 20. Let $\varepsilon_n := \varepsilon > 0$ for $n \in \mathbb{N}_0$, $r \in \mathbb{N}$, $r > 1$,

$$
a_{r^n} = 2, \qquad n \in \mathbb{N}_0,
$$

$$
a_k = 1, \qquad k \notin \{r^n : n \in \mathbb{N}_0\}.
$$

It is easy to check that

$$
\liminf_{n\to\infty} \frac{|a_{n+1}|\varepsilon_n}{\varepsilon_{n+1}} = 1
$$

and

$$
\sum_{n=0}^{\infty} \frac{1}{|a_0 \ldots a_n|} = \infty, \qquad \lim_{n\to\infty} |a_0 \ldots a_{n-1}| = \infty.
$$

This means that condition (c) of Theorem 109 holds. Consequently, difference equation (6.15) is $(\varepsilon_n)_{n\in\mathbb{N}_0}$-nonstable in this situation.

From Theorem 109 the following corollary can be derived very easily:

Corollary 13. *Suppose that (6.17) holds and there exists $a \in \mathbb{K}$, $a \neq 0$, such that*

$$\lim_{n \to \infty} a_n = a. \tag{6.31}$$

Then for each $\widehat{y} \in X$ there is a sequence $(y_n)_{n \in \mathbb{N}_0}$ in X such that $y_0 = \widehat{y}$,

$$\|y_{n+1} - a_n y_n - b_n\| \leq \varepsilon_n, \qquad n \in \mathbb{N}_0, \tag{6.32}$$

and, for arbitrary sequence $(x_n)_{n \in \mathbb{N}_0}$ in X with

$$x_{n+1} = a_n x_n + b_n, \qquad n \in \mathbb{N}_0, \tag{6.33}$$

we have

$$\sup_{n \in \mathbb{N}_0} \frac{\|y_n - x_n\|}{\varepsilon_{n+k}} = \sup_{n \in \mathbb{N}_0} \frac{\|y_{n+k} - x_{n+k}\|}{\varepsilon_n} = \infty, \qquad k \in \mathbb{N}_0. \tag{6.34}$$

Proof. Note that (6.31) yields

$$\lim_{n \to \infty} \frac{\varepsilon_{n+1}}{\varepsilon_n} = |a|, \tag{6.35}$$

Fix $\widehat{y} \in X$. According to Theorem 109, there is a sequence $(y_n)_{n \in \mathbb{N}_0}$ in X such that $y_0 = \widehat{y}$, (6.32) holds and, for each sequence $(x_n)_{n \in \mathbb{N}_0}$ in X satisfying (6.33),

$$\sup_{n \in \mathbb{N}} \frac{\|y_n - x_n\|}{\varepsilon_{n-1}} = \infty. \tag{6.36}$$

Take $k \in \mathbb{N}$. Then, by (6.35),

$$\lim_{n \to \infty} \frac{\varepsilon_{n+k}}{\varepsilon_n} = |a|^k, \qquad \lim_{n \to \infty} \frac{\varepsilon_n}{\varepsilon_{n+k}} = |a|^{-k},$$

whence we deduce that (6.36) implies (6.34). $\qquad\square$

The next corollary is a simplified version of Corollary 13.

Corollary 14. *Let $a \in \mathbb{K}$, $|a| = 1$, and $\delta > 0$. Then for each $\widehat{y} \in X$ there exists a sequence $(y_n)_{n \in \mathbb{N}_0}$ in X such that $y_0 = \widehat{y}$,*

$$\|y_{n+1} - a y_n - b_n\| \leq \delta, \qquad n \in \mathbb{N}_0, \tag{6.37}$$

and, for arbitrary sequence $(x_n)_{n \in \mathbb{N}_0}$ in X satisfying the difference equation

$$x_{n+1} = a x_n + b_n, \qquad n \in \mathbb{N}_0, \tag{6.38}$$

we have

$$\sup_{n \in \mathbb{N}_0} \|y_n - x_n\| = \infty. \tag{6.39}$$

We end this section with a proposition, which shows that such sequences $(y_n)_{n\in\mathbb{N}_0}$, proving nonstability of (6.33), can be constructed in a more general situation. The proof of it has been based on some reasonings from [3].

Proposition 11. *Assume that $a : X \to X$ is a linear isometry, i.e., it is linear and*

$$\|a(x)\| = \|x\|, \qquad x \in X.$$

Then, for each $y \in X$ and each real number $\delta_0 > 0$, there exists an unbounded sequence $(y_n)_{n\in\mathbb{N}_0}$ in X with $y_0 = y$ and

$$\|y_{n+1} - a(y_n) - b_n\| = \delta_0, \qquad n \in \mathbb{N}_0, \tag{6.40}$$

such that

$$\sup_{n\in\mathbb{N}_0} \|y_n - x_n\| = \infty \tag{6.41}$$

for every sequence $(x_n)_{n\in\mathbb{N}_0}$ in X satisfying the difference equation

$$x_{n+1} = a(x_n) + b_n, \qquad n \in \mathbb{N}_0. \tag{6.42}$$

Proof. Fix $y \in X$ and a real number $\delta_0 > 0$. Let $u \in X$, $\|u\| = 1$,

$$\varepsilon := \begin{cases} 1, & \sup_{n\in\mathbb{N}_0} \left\| \sum_{k=1}^n a^{n-k}(b_{k-1}) + n\delta_0 a^n(u) \right\| = \infty; \\ -1, & \text{otherwise,} \end{cases}$$

and $(y_n)_{n\in\mathbb{N}_0}$ be given by

$$y_{n+1} = a(y_n) + b_n + \varepsilon\delta_0 a^{n+1}(u), \qquad n \in \mathbb{N}_0.$$

Clearly,

$$\|y_{n+1} - a(y_n) - b_n\| = \delta_0, \qquad n \in \mathbb{N}_0.$$

Take $x_0 \in X$ and define $(x_n)_{n\in\mathbb{N}_0}$ by (6.42). Then, by induction on n, we obtain that

$$y_n - x_n = a^n(y_0 - x_0) + n\delta_0 \varepsilon a^n(u), \qquad n \in \mathbb{N}_0,$$

and consequently

$$\|y_n - x_n\| = \|a^n(y_0 - x_0 + n\delta_0 \varepsilon u)\| = \|y_0 - x_0 + n\delta_0 \varepsilon u\|, \qquad n \in \mathbb{N}_0.$$

Since

$$\|y_0 - x_0 + n\delta_0 \varepsilon u\| \geq \left| n\delta_0 \|u\| - \|y_0 - x_0\| \right|, \qquad n \in \mathbb{N}_0,$$

it is easily seen that

$$\lim_{n\to\infty} \|y_n - x_n\| = \infty.$$

To complete the proof observe that, again by induction, for every $n \in \mathbb{N}$ we get

$$y_n = a^n(y_0) + \sum_{k=1}^n a^{n-k}(b_{k-1}) + n\varepsilon\delta_0 a^n(u),$$

whence, in the case $\varepsilon = 1$,

$$\|y_n\| \geq \left\| \sum_{k=1}^n a^{n-k}(b_{k-1}) + n\delta_0 a^n(u) \right\| - \|y_0\|,$$

and, in the case $\varepsilon = -1$,

$$\|y_n\| = \left\| a^n y_0 + \sum_{k=1}^n a^{n-k}(b_{k-1}) - n\delta_0 a^n(u) \right\|$$

$$\geq 2n\delta_0 - \left\| \sum_{k=1}^n a^{n-k}(b_{k-1}) + n\delta_0 a^n(u) \right\| - \|y_0\|,$$

which, in either case, means that $(y_n)_{n\in\mathbb{N}_0}$ is unbounded. $\qquad\square$

The subsequent proposition will be useful in the next section.

Corollary 15. *Assume that $I : X \to X$ is a bijective linear isometry and $\alpha_0 \in \mathbb{K}$, $|\alpha_0| = 1$. Then for each $y_0 \in X$ there exists an unbounded sequence $(y_n)_{n\in\mathbb{N}_0}$ in X, satisfying the inequality*

$$\|I(y_{n+1}) - \alpha_0 y_n - b_n\| \leq \delta, \qquad n \in \mathbb{N}_0, \tag{6.43}$$

such that (6.41) holds for arbitrary sequence $(x_n)_{n\in\mathbb{N}_0}$ in X, satisfying the difference equation

$$I(x_{n+1}) = \alpha_0 x_n + b_n, \qquad n \in \mathbb{N}_0. \tag{6.44}$$

Proof. It is enough to observe that (6.43) is equivalent to the inequality

$$\|y_{n+1} - \alpha_0 I^{-1}(y_n) - I^{-1}(b_n)\| \leq \delta, \qquad n \in \mathbb{N}_0,$$

and use Proposition 11 with $a(x) = \alpha_0 I^{-1}(x)$ for $x \in X$ and b_n replaced by $I^{-1}(b_n)$ for each $n \in \mathbb{N}_0$. $\qquad\square$

Let us yet state a simplified version of Proposition 11, corresponding to Corollary 14.

Corollary 16. *Let B be a normed algebra, $a \in B$, $\|a\| = 1$, and $\delta > 0$. Then for each*

$y_0 \in B$ *there exists an unbounded sequence* $(y_n)_{n \in \mathbb{N}_0}$ *in B such that*

$$\|y_{n+1} - ay_n - b_n\| \leq \delta, \qquad n \in \mathbb{N}_0, \tag{6.45}$$

and, for arbitrary sequence $(x_n)_{n \in \mathbb{N}_0}$ *in B satisfying the difference equation*

$$x_{n+1} = ax_n + b_n, \qquad n \in \mathbb{N}_0, \tag{6.46}$$

we have

$$\sup_{n \in \mathbb{N}_0} \|y_n - x_n\| = \infty. \tag{6.47}$$

4. Linear difference equation of a higher order

In this section we discuss the nonstability of the difference equation

$$I^m(x_{n+m}) + \sum_{i=0}^{m-1} a_i I^i(x_{n+i}) = b_n, \qquad n \in S, \tag{6.48}$$

where $S \in \{\mathbb{N}_0, \mathbb{Z}\}$, $m \in \mathbb{N}$, $a_0, ..., a_{m-1} \in \mathbb{K}$, $I : X \to X$ is a bijective linear isometry, and $(b_n)_{n \in \mathbb{N}_0}$ is a given sequence in X. Equation (6.48) is a natural generalization of the linear difference equation of a higher order with constant coefficients, i.e., of the equation

$$x_{n+m} + \sum_{i=0}^{m-1} a_i x_{n+i} = b_n, \qquad n \in \mathbb{N}_0, \tag{6.49}$$

In what follows $r_1, r_2, ..., r_m \in \mathbb{C}$ stand for the roots of the equation

$$r^m + \sum_{i=0}^{m-1} a_i r^i = 0, \tag{6.50}$$

which is the characteristic equation of (6.48).

The following result generalizes [3, Theorem 4] and some results in [6], which also concern (6.48), but with $I(x) = x$ for all $x \in X$.

Theorem 110. *Suppose* $r_1, r_2, ..., r_m \in \mathbb{K}$ *and* $|r_i| = 1$ *for some* $i \in \{1, ..., m\}$. *Then for any* $\delta > 0$ *there exists an unbounded sequence* $(y_n)_{n \in \mathbb{N}_0}$ *in X, satisfying the inequality*

$$\left\| I^m(y_{n+m}) + \sum_{i=0}^{m-1} a_i I^i(y_{n+i}) - b_n \right\| \leq \delta, \qquad n \in \mathbb{N}_0, \tag{6.51}$$

such that for every sequence $(x_n)_{n \in \mathbb{N}_0}$ *in X, fulfilling the linear difference equation*

(6.48), *we have*

$$\sup_{n \in \mathbb{N}_0} \|y_n - x_n\| = \infty. \tag{6.52}$$

Proof. The proof is analogous as for [3, Theorem 4]. It is by induction on m.

The case $m = 1$ is true in view of Corollary 15.

For $m \geq 2$, without loss of generality, we can assume that $|r_1| = 1$. From Corollary 15 it follows that there exists an unbounded sequence $(\widetilde{y}_n)_{n \in \mathbb{N}_0}$ in X, satisfying the inequality

$$\|I(\widetilde{y}_{n+1}) - r_1 \widetilde{y}_n - b_n\| \leq \delta, \qquad n \in \mathbb{N}_0, \tag{6.53}$$

such that for every sequence $(\widetilde{x}_n)_{n \in \mathbb{N}_0}$ with

$$I(\widetilde{x}_{n+1}) = r_1 \widetilde{x}_n + b_n, \qquad n \in \mathbb{N}_0, \tag{6.54}$$

we have

$$\sup_{n \in \mathbb{N}_0} \|\widetilde{y}_n - \widetilde{x}_n\| = \infty. \tag{6.55}$$

Further, there exists a sequence $(y_n)_{n \in \mathbb{N}_0}$ in X such that

$$\widetilde{y}_n = I^{m-1}(y_{n+m-1}) + (-1)(r_2 + \ldots + r_m) I^{m-2}(y_{n+m-2}) \tag{6.56}$$
$$+ \ldots + (-1)^{m-1} r_2 \ldots r_m y_n, \qquad n \in \mathbb{N}_0;$$

it suffices to take $y_0 = \ldots = y_{m-2} = 0$, and then determine $(y_n)_{n \in \mathbb{N}_0}$ step by step, by the formula

$$y_{n+m-1} = I^{1-m}((r_2 + \ldots + r_m) I^{m-2}(y_{n+m-2})$$
$$- \ldots - (-1)^{m-1} r_2 \ldots r_m y_n + \widetilde{y}_n), \qquad n \in \mathbb{N}_0.$$

Inequality (6.53) implies that the sequence $(y_n)_{n \in \mathbb{N}_0}$ satisfies the inequality

$$\|I^m(y_{n+m}) + (-1)(r_1 + \ldots + r_m) I^{m-1}(y_{n+m-1}) \tag{6.57}$$
$$+ \ldots + (-1)^m r_1 \ldots r_m y_n - b_n\| \leq \delta, \qquad n \in \mathbb{N}_0,$$

which is (6.51).

Let now $(x_n)_{n \in \mathbb{N}_0}$ be an arbitrary sequence defined by (6.48) and $(\widetilde{x}_n)_{n \in \mathbb{N}_0}$ be the sequence given by

$$\widetilde{x}_n = I^{m-1}(x_{n+m-1}) + (-1)(r_2 + \ldots + r_m) I^{m-2}(x_{n+m-2}) \tag{6.58}$$
$$+ \ldots + (-1)^{m-1} r_2 \ldots r_m x_n, \qquad n \in \mathbb{N}_0.$$

Then (6.54) and (6.55) holds.

We are to prove that

$$\sup_{n \in \mathbb{N}_0} \|x_n - y_n\| = \infty.$$

Suppose the contrary. Then there exists $M > 0$ such that

$$\|y_n - x_n\| \le M, \qquad n \in \mathbb{N}_0,$$

whence (6.56) and (6.58) imply that

$$\|\widetilde{y}_n - \widetilde{x}_n\| \le \|y_{n+m-1} - x_{n+m-1}\| + |r_2 + \ldots + r_m| \|y_{n+m-2} - x_{n+m-2}\|$$
$$+ \ldots + |r_2 \ldots r_m| \|y_n - x_n\|$$
$$\le (1 + |r_2 + \ldots + r_m| + \ldots + |r_2 \ldots r_m|)M$$

for every $n \in \mathbb{N}_0$, which contradicts (6.55).

To end the proof observe that, by (6.56), $(y_n)_{n \in \mathbb{N}_0}$ is unbounded. $\qquad \square$

The next example (see [3, Remark 1]) shows how we can construct such sequence $(y_n)_{n \in \mathbb{N}_0}$ in a quite simple situation.

Remark 29. Let $\mathbb{K} = \mathbb{R}$. Consider the linear difference equation (6.48) with $m = 2$ and $I(x) \equiv x$, i.e., the equation

$$x_{n+2} = -x_n, \qquad n \in \mathbb{N}_0. \tag{6.59}$$

The characteristic equation of it is of the form

$$z^2 = -1$$

and has no real roots.

Let $\delta > 0$, $u \in X$, $\|u\| = 1$. Take $y_0 \in X$ and define a sequence $(y_n)_{n \in \mathbb{N}_0}$ in X by

$$y_{n+2} = -y_n + (-1)^{[\frac{n}{2}]} \delta u, \qquad n \in \mathbb{N}_0,$$

($[s]$ denotes the integer part of $s \in \mathbb{R}$, i.e., the biggest integer not greater than s). Clearly,

$$\|y_{n+2} + y_n\| \le \delta, \qquad k \in \mathbb{N}_0,$$

and

$$y_{2(k+1)} = -y_{2k} + (-1)^k \delta u, \qquad k \in \mathbb{N}_0. \tag{6.60}$$

We show by induction on k that

$$y_{2k} = (-1)^k y_0 + (-1)^{k-1} k \delta u \tag{6.61}$$

for every $k \in \mathbb{N}$.

The case $k = 1$ is trivial. So, take $j \in \mathbb{N}$ and assume that (6.61) holds for $k = j$. Then, (6.60) implies that

$$
\begin{aligned}
y_{2(j+1)} &= -y_{2j} + (-1)^j \delta u \\
&= -\left((-1)^j y_0 + (-1)^{j-1} j \delta u\right) + (-1)^j \delta u \\
&= (-1)^{j+1} y_0 + (-1)^j (j + 1) \delta u,
\end{aligned}
$$

which completes the proof of (6.61).

Let $(x_n)_{n \in \mathbb{N}_0}$ be a sequence in X satisfying (6.59). Then

$$
\begin{aligned}
\|y_{2k} - x_{2k}\| &= \|(-1)^k y_0 + (-1)^{k-1} k \delta u - (-1)^k x_0\| \\
&\geq \left| \|(-1)^k (y_0 - x_0)\| - \|(-1)^{k-1} k \delta u\| \right| \\
&= \left| \|y_0 - x_0\| - k\delta \right|
\end{aligned}
$$

for every $k \in \mathbb{N}$, whence

$$
\sup_{n \in \mathbb{N}_0} \|y_n - x_n\| = \infty.
$$

This example shows how to find a sequence $(y_n)_{n \in S}$ proving nonstability of difference equation (6.48), in some particular cases with $\mathbb{K} = \mathbb{R}$.

Now, we prove that in a general situation, for $\mathbb{K} = \mathbb{R}$, the assumption that all the roots of the characteristic equation (6.50) are real is not necessary in Theorem 110. Therefore the next theorem (see [6, Theorem 4]) complements Theorem 110.

Theorem 111. *Let $S \in \{\mathbb{N}_0, \mathbb{Z}\}$ and $|r_j| = 1$ for some $j \in \{1, \ldots, m\}$. Then, for any $\delta > 0$, there exists a sequence $(y_n)_{n \in S}$ in X, satisfying the inequality*

$$
\left\| I^m(y_{n+m}) + \sum_{i=0}^{m-1} a_i I^i(y_{n+i}) - b_n \right\| \leq \delta, \qquad n \in S, \tag{6.62}
$$

such that

$$
\sup_{n \in S} \|y_n - x_n\| = \infty \tag{6.63}
$$

for every sequence $(x_n)_{n \in S}$ in X, fulfilling the difference equation

$$
I^m(x_{n+m}) + \sum_{i=0}^{m-1} a_i I^i(x_{n+i}) = b_n, \qquad n \in S. \tag{6.64}
$$

Moreover, if $r_1, \ldots, r_m \in \mathbb{K}$ or there is a bounded sequence $(x_n)_{n \in S}$ in X fulfilling (6.64), then $(y_n)_{n \in S}$ can be chosen unbounded.

Proof. First consider the case when $r_i \in \mathbb{K}$ for $i = 1, \ldots, m$. Clearly, the case $S = \mathbb{N}_0$ follows directly from Theorem 110. So assume that $S = \mathbb{Z}$. Let $(y_n)_{n \in \mathbb{N}_0}$ be an unbounded sequence in X, the existence of which is guaranteed by Theorem 110.

Since $|r_j| = 1$ for some $j \in \{1, \ldots, m\}$, there is $i \in \{0, \ldots, m - 1\}$ with $a_i \neq 0$. Let

$$s := \min \{i \in \{1, \ldots, m - 1\} : a_i \neq 0\}$$

and, in the case $s < m - 1$, write

$$y_{-n} := I^{-s} \left(\frac{-1}{a_s} \left(I^m(y_{m-n-s}) + \sum_{i=s+1}^{m-1} a_i I^i(y_{i-n-s}) - b_{-n-s} \right) \right), \qquad n \in \mathbb{N};$$

if $s = m - 1$, then we write

$$y_{-n} := I^{1-m} \left(\frac{-1}{a_{m-1}} \left(I^m(y_{-n+1}) - b_{-n-m+1} \right) \right), \qquad n \in \mathbb{N}.$$

Clearly, in either case, the sequence $(y_n)_{n \in \mathbb{Z}}$ is unbounded,

$$\sup_{n \in \mathbb{Z}} \left\| I^m(y_{n+m}) + \sum_{i=0}^{m-1} a_i I^i(y_{n+i}) - b_n \right\| \leq \delta,$$

and

$$\sup_{n \in \mathbb{Z}} \|y_n - x_n\| = \infty$$

for every sequence $(x_n)_{n \in \mathbb{Z}}$ in X, fulfilling recurrence (6.64) with $S = \mathbb{Z}$.

It remains to consider the case where $m > 1$, $\mathbb{K} = \mathbb{R}$ and $r_1 \notin \mathbb{R}$. Let $\mathfrak{C}(X)$ denote, as before, the complexification of X (with the Taylor norm $\| \cdot \|_T$) and (as before) define $\pi_i : X^2 \to X$ by: $\pi_i(x_1, x_2) := x_i$ for $x_1, x_2 \in X$, $i = 1, 2$.

Define $\widehat{I} : \mathfrak{C}(X) \to \mathfrak{C}(X)$ and a sequence $(\widehat{b}_n)_{n \in \mathbb{N}_0}$ in $\mathfrak{C}(X)$ by

$$\widehat{I}(x, y) = (I(x), I(y)), \qquad (x, y) \in \mathfrak{C}(X),$$

$$\widehat{b}_n := (b_n, b_n), \qquad n \in S.$$

Then it is easy to see that $\widehat{I}$ is a bijective linear isometry, and therefore, by Theorem 110, there is an unbounded sequence $(\widehat{y}_n)_{n \in \mathbb{N}_0}$ in $\mathfrak{C}(X)$ with

$$\sup_{n \in \mathbb{N}_0} \left\| \widehat{I}^m(\widehat{y}_{n+m}) + \sum_{i=0}^{m-1} a_i \widehat{I}^i(\widehat{y}_{n+i}) - \widehat{b}_n \right\|_T \leq \delta$$

and

$$\sup_{n \in \mathbb{N}_0} \|\widehat{y}_n - \widehat{x}_n\|_T = \infty \tag{6.65}$$

for every sequence $(\widehat{x}_n)_{n\in\mathbb{N}_0}$ in $\mathfrak{C}(X)$, given by

$$\widehat{I}^m(\widehat{x}_{n+m}) + \sum_{i=0}^{m-1} a_i\widehat{I}^i(\widehat{x}_{n+i}) = \widehat{b}_n. \tag{6.66}$$

Next, if $S = \mathbb{Z}$, then analogously as before, in the case $s < m - 1$, we write

$$\widehat{y}_{-n} := \widehat{I}^{-s}\left(\frac{-1}{a_s}\left(\widehat{I}^m(\widehat{y}_{m-n-s}) + \sum_{i=s+1}^{m-1} a_i\widehat{I}^i(\widehat{y}_{i-n-s}) - \widehat{b}_{-n-s}\right)\right), \qquad n \in \mathbb{N};$$

if $s = m - 1$, then we write

$$\widehat{y}_{-n} := \widehat{I}^{1-m}\left(\frac{-1}{a_{m-1}}\left(\widehat{I}^m(\widehat{y}_{-n+1}) - \widehat{b}_{-n-m+1}\right)\right), \qquad n \in \mathbb{N}.$$

Clearly

$$\sup_{n\in\mathbb{Z}}\left\|\widehat{I}^m(\widehat{y}_{n+m}) + \sum_{i=0}^{m-1} a_i\widehat{I}^i(\widehat{y}_{n+i}) - \widehat{b}_n\right\|_T \le \delta$$

and

$$\sup_{n\in\mathbb{Z}}\|\widehat{y}_n - \widehat{x}_n\|_T = \infty$$

for every sequence $(\widehat{x}_n)_{n\in\mathbb{Z}}$ in X^2, satisfying (6.66) for $n \in \mathbb{Z}$.

We show yet that there is $j \in \{1, 2\}$ such that the sequence $(\pi_j(\widehat{y}_n) - x_n)_{n\in S}$ is unbounded for every sequence $(x_n)_{n\in S}$ in X, fulfilling (6.64),

For the proof by contradiction suppose that there are sequences $(x_n)_{n\in S}$ and $(x'_n)_{n\in S}$ in X, fulfilling difference equation (6.64), such that the sequences

$$(\pi_1(\widehat{y}_n) - x_n)_{n\in S}, \qquad (\pi_2(\widehat{y}_n) - x'_n)_{n\in S}$$

are bounded. Write

$$\widehat{x}_n := (x_n, x'_n), \qquad n \in S.$$

It is easily seen that (6.66) holds and

$$\|\widehat{y}_n - \widehat{x}_n\|_T \le \|\pi_1(\widehat{y}_n) - x_n\| + \|\pi_2(\widehat{y}_n) - x'_n\|, \qquad n \in S,$$

whence the sequence $(\widehat{y}_n - \widehat{x}_n)_{n\in S}$ is bounded. This is a contradiction to (6.65).

Now, it is easily seen that it is enough to take $y_n := \pi_j(\widehat{y}_n)$ for $n \in S$.

To complete the proof observe that, if there exists a bounded sequence $(x_n)_{n\in S}$ in X such that (6.64) holds, then $(y_n)_{n\in S}$ must be unbounded, because

$$\sup_{n\in S}\|y_n - x_n\| = \infty.$$

$\square$

5. Linear functional equation of the first order

J.A. Baker [1] discussed stability for linear functional equations of the following form:

$$\varphi(x) = g(x)\varphi(f(x)) + h(x). \tag{6.67}$$

The following was obtained:

Theorem 112. *Let X be complete, S be a nonempty set, $f : S \to S$, $h : S \to X$, $g : S \to \mathbb{K}$,*

$$|g(x)| \le \lambda, \qquad x \in S,$$

and $0 \le \lambda < 1$. Suppose that $\varphi_s : S \to X$ satisfies

$$\delta := \sup_{x \in S} \|\varphi_s(x) - g(x)\varphi_s(f(x)) - h(x)\| < \infty. \tag{6.68}$$

Then there exists a unique solution $\varphi : S \to X$ of equation (6.67) with

$$\sup_{x \in S} \|\varphi_s(x) - \varphi(x)\| \le \frac{\delta}{1 - \lambda}. \tag{6.69}$$

Moreover, he also proved a similar result in a Banach algebra.

It is easily seen that Theorem 112 is applicable to linear equations of the form

$$\varphi(f(x)) = a(x)\varphi(x) + h(x). \tag{6.70}$$

because, under suitable assumptions on a and f, we can rewrite (6.70) as

$$\varphi(x) = (a(x))^{-1}\varphi(f(x)) - (a(x))^{-1}h(x), \tag{6.71}$$

or

$$\varphi(x) = a(f^{-1}(x))\varphi(f^{-1}(x)) + h(f^{-1}(x)). \tag{6.72}$$

Next, S.H. Lee and K.W. Jun [18] considered stability of a particular form of (6.70), i.e., of the equation

$$\varphi(x + p) = k\varphi(x), \tag{6.73}$$

while S.M. Jung [12, 13, 14] discussed stability of the gamma functional equation

$$\varphi(x + 1) = x\varphi(x), \qquad x \in (0, +\infty). \tag{6.74}$$

Later G.H. Kim [15] (see also [16]) generalized Jung's results to the generalized gamma functional equation

$$\varphi(x + p) = a(x)\varphi(x), \qquad x \in (0, +\infty), \tag{6.75}$$

with a suitable $a : (0, \infty) \to (0, \infty)$, and obtained the following two theorems:

Theorem 113. *If $x_0, \delta \in \mathbb{R}_+$, functions $a : (0, \infty) \to (0, \infty)$ and $\varphi_s : (0, \infty) \to \mathbb{R}$ satisfy the inequality*

$$|\varphi_s(x + p) - a(x)\varphi_s(x)| \leq \delta, \qquad x > x_0,$$

and

$$\gamma(x) := \sum_{j=0}^{\infty} \prod_{i=0}^{j} \frac{1}{a(x + pi)} < \infty, \qquad x > x_0,$$

then there exists a unique solution $\varphi : (0, \infty) \to \mathbb{R}$ of equation (6.75) such that

$$|\varphi_s(x) - \varphi(x)| \leq \gamma(x)\delta, \qquad x > x_0.$$

Theorem 114. *If the functions $\varphi_s : (0, \infty) \to \mathbb{R}$ and $\varepsilon, a : (0, \infty) \to (0, \infty)$ satisfy the inequality*

$$|\varphi_s(x + p) - a(x)\varphi_s(x)| \leq \varepsilon(x), \qquad x > x_0,$$

with some $x_0 \in \mathbb{R}_+$, and

$$\Phi(x) := \sum_{j=0}^{\infty} \varepsilon(x + pj) \prod_{i=0}^{j} \frac{1}{a(x + pi)} < \infty, \qquad x > x_0,$$

then there exists a unique solution $\varphi : (0, \infty) \to \mathbb{R}$ of equation (6.75) with

$$|\varphi_s(x) - \varphi(x)| \leq \Phi(x), \qquad x > x_0.$$

Finally, let us mention that T. Trif [21] has proved the following result.

Theorem 115. *Let X be complete and S be a nonempty set. Let $f : S \to S, h : S \to X$, $a : S \to \mathbb{K} \setminus \{0\}$ and $\varepsilon : S \to \mathbb{R}_+$ be given functions such that*

$$\omega(x) := \sum_{k=0}^{\infty} \frac{\varepsilon(f^k(x))}{\prod_{j=0}^{k} |a(f^j(x))|} < \infty, \qquad x \in S. \tag{6.76}$$

If a function $\varphi_s : S \to X$ satisfies

$$\|\varphi_s(f(x)) - a(x)\varphi_s(x) - h(x)\| \leq \varepsilon(x), \qquad x \in S, \tag{6.77}$$

then there exists a unique solution $\varphi : S \to X$ of equation (6.70) with

$$\|\varphi_s(x) - \varphi(x)\| \le \omega(x), \qquad x \in S. \tag{6.78}$$

Let A be a nonempty set, $a : A \to \mathbb{K} \setminus \{0\}$, $f : A \to A$ and $F : A \to X$. In this section we present nonstability results, related to all those stability outcomes mentioned above and concerning the functional equation

$$\varphi(f(x)) = a(x)\varphi(x) + F(x) \tag{6.79}$$

in the class of functions $\varphi : A \to X$. To this end we need the subsequent hypothesis for functions $\varepsilon : A \to \mathbb{R}_+$.

$(\mathcal{H})$ There is $u \in A$ such that the set $Orb(u) := \{f^k(u) : k \in \mathbb{N}_0\}$ has infinitely many elements and

$$\lim_{n \to \infty} \frac{\varepsilon(f^{n+1}(u))}{\varepsilon(f^n(u))|a(f^{n+1}(u))|} = 1. \tag{6.80}$$

We start with a lemma.

Lemma 11. *Assume that $(\mathcal{H})$ holds. Then*

$$f^j(u) \ne f^k(u), \qquad j, k \in \mathbb{N}_0, j \ne k, \tag{6.81}$$

$$A^{-j} \cap A^{-k} = \emptyset, \qquad j, k \in \mathbb{N}_0, j \ne k, \tag{6.82}$$

where $A^0 := Orb(u)$ and

$$A^{-n} := f^{-1}(A^{-n+1}) \setminus A^{-n+1}, \qquad n \in \mathbb{N}.$$

Proof. First we show that (6.81) holds. So, suppose that $f^j(u) = f^k(u)$ for some $j, k \in \mathbb{N}_0$, $j < k$. Then

$$f^{k-j}(f^j(u)) = f^k(u) = f^j(u). \tag{6.83}$$

Clearly, for each $n \in \mathbb{N}$, $n > j$, there are $l, r \in \mathbb{N}_0$ with $r < k - j$ and

$$n - j = (k - j)l + r,$$

whence (6.83) implies that

$$f^n(u) = f^{n-j}(f^j(u)) = f^r(f^{(k-j)l}(f^j(u))) = f^r(f^j(u)).$$

This contradicts the assumption that $Orb(u)$ is infinite. Thus we have proved (6.81).

Next, for the proof of (6.82) suppose that there are $j, k \in \mathbb{N}_0$, $j < k$, with

$$A^{-j} \cap A^{-k} \ne \emptyset.$$

Let $z \in A^{-j} \cap A^{-k}$. Note that $f^j(z), f^k(z) \in Orb(u)$ and therefore

$$f^{k-1}(z) = f^{k-j-1}(f^j(z)) \in Orb(u),$$

which is a contradiction, because $f^{k-1}(z) \in A^{-1}$ and $A^{-1} \cap Orb(u) = \emptyset$. This completes the proof of (6.82).   □

Now, we are in a position to prove the following simplified version of [8, Theorem 3.2].

Proposition 12. *Assume that hypothesis $(\mathcal{H})$ is valid and $\varepsilon : A \to (0, \infty)$. Then there exists a mapping $\varphi : A \to X$ satisfying the inequality*

$$\|\varphi(f(x)) - a(x)\varphi(x)\| \le \varepsilon(x), \qquad x \in A, \tag{6.84}$$

and such that

$$\sup_{x \in A} \frac{\|\varphi(f(x)) - \widetilde{\varphi}(f(x))\|}{\varepsilon(x)} = \infty \tag{6.85}$$

for each solution $\widetilde{\varphi} : A \to X$ of the equation

$$\varphi(f(x)) = a(x)\varphi(x). \tag{6.86}$$

Moreover, if additionally

$$\inf_{n \in \mathbb{N}} \frac{\varepsilon(f^n(u))}{\varepsilon(f^{n+1}(u))} > 0, \qquad \sup_{n \in \mathbb{N}} \frac{\varepsilon(f^n(u))}{\varepsilon(f^{n+1}(u))} < \infty, \tag{6.87}$$

then

$$\sup_{x \in A} \frac{\|\varphi(x) - \widetilde{\varphi}(x)\|}{\varepsilon(f^k(x))} = \infty, \qquad \sup_{x \in A} \frac{\|\varphi(f^k(x)) - \widetilde{\varphi}(f^k(x))\|}{\varepsilon(x)} = \infty \tag{6.88}$$

for each $k \in \mathbb{N}_0$.

Proof. Write

$$A_u := \{y \in A : \exists_{k,m \in \mathbb{N}} \ f^k(y) = f^m(u)\},$$

$\varepsilon_n := \varepsilon(f^n(u))$ and $a_n := -a(f^n(u))$ for $n \in \mathbb{N}_0$. Then (6.17) holds and, by Corollary 12 (with $b_n := 0$ for $n \in \mathbb{N}_0$), there is a sequence $(y_n)_{n \in \mathbb{N}_0}$ in X with

$$\|y_{n+1} + a_n y_n\| \le \varepsilon_n, \qquad n \in \mathbb{N}_0, \tag{6.89}$$

and such that

$$\sup_{n \in \mathbb{N}} \frac{\|y_n - x_n\|}{\varepsilon_{n-1}} = \infty \tag{6.90}$$

for each sequence $(x_n)_{n\in\mathbb{N}_0}$ in X with

$$x_{n+1} + a_n x_n = 0, \qquad n \in \mathbb{N}_0. \tag{6.91}$$

Define the function $\varphi : A \to X$ as follows:

$$\varphi(x) = 0, \qquad x \in A \setminus A_u,$$

$$\varphi(f^n(u)) = y_n, \qquad n \in \mathbb{N}_0,$$

$$\varphi(x) = \frac{1}{a(x)}(\varphi(f(x))), \qquad x \in A^{-n}, n \in \mathbb{N}.$$

In view of Lemma 11, (6.81) and (6.82) are valid, and therefore, the definition is correct. Note that

$$\varphi(f(x)) = a(x)\varphi(x), \qquad x \in A \setminus Orb(u).$$

Further,

$$\varphi(f(f^n(u))) - a(f^n(u))\varphi(f^n(u)) = y_{n+1} + a_n y_n$$

for all $n \in \mathbb{N}_0$, whence we get (6.84) (on account of (6.89)). Take a function $\widetilde{\varphi} : A \to X$ satisfying (6.86). Let

$$x_n := \widetilde{\varphi}(f^n(u)), \qquad n \in \mathbb{N}_0.$$

Then $(x_n)_{n\in\mathbb{N}_0}$ satisfies difference equation (6.91) and consequently (6.90) holds. Moreover,

$$\sup_{x\in A} \frac{\|\varphi(f(x)) - \widetilde{\varphi}(f(x))\|}{\varepsilon(x)} \geq \sup_{n\in\mathbb{N}} \frac{\|y_n - x_n\|}{\varepsilon_{n-1}} = \infty. \tag{6.92}$$

If we assume additionally that (6.87) holds, then

$$\inf_{n\in\mathbb{N}} \frac{\varepsilon(f^n(u))}{\varepsilon(f^{n+k}(u))} > 0, \qquad \inf_{n\in\mathbb{N}} \frac{\varepsilon(f^{n+k}(u))}{\varepsilon(f^n(u))} > 0, \qquad k \in \mathbb{N}, \tag{6.93}$$

and further

$$\frac{\varepsilon(f^{n+k-1}(u))}{\varepsilon(f^n(u))} \frac{\|\varphi(f^{n+k}(u)) - \widetilde{\varphi}(f^{n+k}(u))\|}{\varepsilon(f^{n+k-1}(u))}$$
$$= \frac{\|\varphi(f^{n+k}(u)) - \widetilde{\varphi}(f^{n+k}(u))\|}{\varepsilon(f^n(u))}, \qquad k \in \mathbb{N}_0, n \in \mathbb{N},$$

and

$$\frac{\varepsilon(f^{n-1}(u))}{\varepsilon(f^{n+k}(u))} \frac{\|\varphi(f^n(u)) - \widetilde{\varphi}(f^n(u))\|}{\varepsilon(f^{n-1}(u))}$$

$$= \frac{\|\varphi(f^n(u)) - \widetilde{\varphi}(f^n(u))\|}{\varepsilon(f^{n+k}(u))}, \qquad k \in \mathbb{N}_0, n \in \mathbb{N}.$$

Hence, from (6.92) and (6.93), we deduce (6.88).   □

Now, we have the following (cf. [8, Theorem 3.2]):

Theorem 116. *Assume that hypothesis ($\mathcal{H}$) is valid, equation (6.79) has a solution $\phi_0 : A \to X$, and $\varepsilon : A \to (0, \infty)$. Then there exists a mapping $\varphi : A \to X$ satisfying inequality*

$$\|\varphi(f(x)) - a(x)\varphi(x) - F(x)\| \leq \varepsilon(x), \qquad x \in A, \tag{6.94}$$

and such that, for each solution $\widetilde{\varphi} : A \to X$ of equation (6.79),

$$\sup_{x \in A} \frac{\|\varphi(f(x)) - \widetilde{\varphi}(f(x))\|}{\varepsilon(x)} = \infty; \tag{6.95}$$

moreover, if additionally

$$\inf_{n \in \mathbb{N}} \frac{\varepsilon(f^n(u))}{\varepsilon(f^{n+1}(u))} > 0, \qquad \sup_{n \in \mathbb{N}} \frac{\varepsilon(f^n(u))}{\varepsilon(f^{n+1}(u))} < \infty, \tag{6.96}$$

then

$$\sup_{x \in A} \frac{\|\varphi(x) - \widetilde{\varphi}(x)\|}{\varepsilon(f^k(x))} = \infty, \qquad \sup_{x \in A} \frac{\|\varphi(f^k(x)) - \widetilde{\varphi}(f^k(x))\|}{\varepsilon(x)} = \infty \tag{6.97}$$

for each $k \in \mathbb{N}_0$.

Proof. In view of Proposition 12, there exists a mapping $\varphi : A \to X$ satisfying inequality (6.84) and such that, for each solution $\widetilde{\varphi} : A \to X$ of equation (6.86), inequality (6.85) is valid.

Write $\varphi_0 := \varphi + \phi_0$. Then, by (6.84),

$$\left\|\varphi_0(f(x)) - a(x)\varphi_0(x) - F(x)\right\| = \|\varphi(f(x))) - a(x)\varphi(x)\|$$

$$\leq \varepsilon(x), \qquad x \in A.$$

Let $\widetilde{\varphi}_0 : A \to X$ be a solution of equation (6.79). Then $\widetilde{\varphi} := \widetilde{\varphi}_0 - \phi_0$ fulfils equation (6.86) and, by (6.85),

$$\sup_{x \in A} \frac{\|\varphi_0(f(x)) - \widetilde{\varphi}_0(f(x))\|}{\varepsilon(x)} = \sup_{x \in A} \frac{\|\varphi(f(x)) - \widetilde{\varphi}(f(x))\|}{\varepsilon(x)} = \infty.$$

Analogously we show that (6.96) implies (6.97) for each $k \in \mathbb{N}_0$.   □

A simple consequence of Theorem 116 is the subsequent corollary (see [8, Corollary 3.1]).

Corollary 17. *Assume that equation* (6.79) *has a solution* $\phi_0 : A \to X$. *Let* ε *be a positive real number and suppose that there exists* $u \in A$ *such that the set* $Orb(u)$ *has infinitely many elements and*

$$\lim_{n \to \infty} |a(f^n(u))| = 1.$$

Then there exists a function $\psi : A \to X$ *such that*

$$\sup_{x \in A} \|\psi(f(x)) - a(x)\psi(x) - F(x)\| \le \varepsilon$$

and, for every solution $\varphi : A \to X$ *of* (6.79),

$$\sup_{x \in f^k(A)} \|\psi(x) - \varphi(x)\| = \infty, \qquad k \in \mathbb{N}_0.$$

6. Linear functional equation of a higher order

Let $m \in \mathbb{N}$, A be a nonempty set, $f : A \to A$, $I : X \to X$ be a bijective linear isometry, $a_0, \ldots, a_{m-1} \in \mathbb{K}$ and $F : A \to X$. In this section we investigate nonstability of the linear functional equation of a high order

$$I^m(\varphi(f^m(x))) + \sum_{j=0}^{m-1} a_j I^j(\varphi(f^j(x))) = F(x), \tag{6.98}$$

in the class of functions $\varphi : A \to X$. We present the outcomes from [7], slightly generalized and/or modified.

They correspond to the notion of stability described in the following definition.

Definition 14. Let $S \subset A$ and $\mathcal{D} \subset X^A$ be nonempty. We say that functional equation (6.98) is Ulam nonstable on the set S, in the class of functions $\mathcal{D}$, provided there is a function $\gamma \in \mathcal{D}$ such that

$$\sup_{x \in S} \left\| I^m(\gamma(f^m(x))) + \sum_{j=0}^{m-1} a_j I^j(\gamma(f^j(x))) - F(x) \right\| < \infty, \tag{6.99}$$

and there does not exist any solution $\varphi \in \mathcal{D}$ of (6.98) with

$$\sup_{x \in S} \|\gamma(x) - \varphi(x)\| < \infty;$$

if $S = A$, then, for simplicity, we omit the part "on the set S"; if $\mathcal{D} = X^A$, then we omit the part "in the class of functions $\mathcal{D}$".

It makes sense to introduce the class $\mathcal{D}$ in Definition (14), because the existence, uniqueness, and behavior of solutions to (6.98) strictly depend on the class of functions in which we investigate them (see, e.g., [17, 0.0B]).

From now on, $r_1, \ldots, r_m \in \mathbb{C}$ stand for the roots of the characteristic equation of (6.98), i.e., of the equation

$$r^m + \sum_{j=0}^{m-1} a_j r^j = 0. \tag{6.100}$$

Remark 30. If $m > 1$, then $b_0, \ldots, b_{m-2}$ denote the unique complex numbers with

$$z^m + \sum_{j=0}^{m-1} a_j z^j = (z - r_1)\left(z^{m-1} + \sum_{j=0}^{m-2} b_j z^j\right), \qquad z \in \mathbb{C}.$$

It is easily seen that $a_{m-1} = b_{m-2} - r_1$, $a_0 = -r_1 b_0$ and, for $m > 3$, $a_j = -r_1 b_j + b_{j-1}$ for $j = 1, \ldots, m - 2$. Moreover, if $r_1, a_0, \ldots, a_{m-1} \in \mathbb{R}$, then $b_0, \ldots, b_{m-2} \in \mathbb{R}$.

First, we prove some auxiliary lemmas concerning the equation

$$I^m(\varphi(f^m(x))) + \sum_{j=0}^{m-1} a_j I^j(\varphi(f^j(x))) = 0. \tag{6.101}$$

We start with a result that is our main tool for investigation of stability of (6.101).

Lemma 12. *Let $r_1 \in \mathbb{K}$, $m > 1$, $T_i \subset A$ be nonempty for $i = 1, 2$, $\psi_0, \psi : A \to X$,*

$$\sup_{x \in T_1} \|I(\psi_0(f(x))) - r_1 \psi_0(x)\| =: \delta < \infty, \tag{6.102}$$

and

$$\sup_{x \in T_i} \left\| I^{m-1}(\psi(f^{m-1}(x))) + \sum_{j=0}^{m-2} b_j I^j(\psi(f^j(x))) - \psi_0(x) \right\| =: \delta_i < \infty \tag{6.103}$$

for $i = 1, 2$. Then the following three conclusions are valid:

(i) If $T_1 \cap f^{-1}(T_1) \neq \emptyset$, then

$$\sup_{x \in T_1 \cap f^{-1}(T_1)} \left\| I^m(\psi(f^m(x))) + \sum_{j=0}^{m-1} a_j I^j(\psi(f^j(x))) \right\| \leq \delta + (1 + |r_1|)\delta_1. \tag{6.104}$$

(ii) If ψ_0 is unbounded on a nonempty $D \subset T_1 \cup T_2$, then ψ is unbounded on the set

$$D_0 := \bigcup_{i=0}^{m-1} f^i(D).$$

(iii) If equation (6.101) has a solution $\varphi : A \to X$ with $\sup_{x \in T_0} \|\psi(x) - \varphi(x)\| < \infty$, where

$$T_0 := \bigcup_{i=0}^{m-1} f^i(T_2),$$

then the functional equation

$$I(\widehat{\eta}(f(x))) = r_1 \widehat{\eta}(x) \tag{6.105}$$

has a solution $\widehat{\eta} : A \to X$ with

$$\sup_{x \in T_2} \|\psi_0(x) - \widehat{\eta}(x)\| < \infty. \tag{6.106}$$

Proof. Note that, according to (6.102), (6.103) and Remark 30,

$$\left\| I^m(\psi(f^m(x))) + \sum_{j=0}^{m-1} a_j I^j(\psi(f^j(x))) \right\|$$

$$\leq \left\| I^{m-1}(\psi(f^{m-1}(f(x)))) + \sum_{j=0}^{m-2} b_j I^j(\psi(f^j(f(x)))) - \psi_0(f(x)) \right\|$$

$$+ |r_1| \left\| I^{m-1}(\psi(f^{m-1}(x))) + \sum_{j=0}^{m-2} b_j I^j(\psi(f^j(x))) - \psi_0(x) \right\|$$

$$+ \left\| I(\psi_0(f(x))) - r_1 \psi_0(x) \right\|$$

$$\leq (1 + |r_1|)\delta_1 + \delta, \qquad x \in T_1 \cap f^{-1}(T_1).$$

Further, if ψ_0 is unbounded on a set $D \subset T_1 \cup T_2$, then (6.103) implies that ψ is unbounded on D_0.

Assume that $\varphi : A \to X$ is a solution to (6.101) such that

$$\sup_{x \in T_0} \|\psi(x) - \varphi(x)\| =: M < \infty$$

and write

$$\widehat{\eta}(x) := I^{m-1}(\varphi(f^{m-1}(x))) + \sum_{j=0}^{m-2} b_j I^j(\varphi(f^j(x))), \qquad x \in A. \tag{6.107}$$

Then $\widehat{\eta} : A \to X$ satisfies equation (6.105) (cf. Remark 30) and (6.103) yields

$$\|\psi_0(x) - \widehat{\eta}(x)\| \leq \left\| I^{m-1}(\psi(f^{m-1}(x))) + \sum_{j=0}^{m-2} b_j I^j(\psi(f^j(x))) - \psi_0(x) \right\|$$

$$+ \left\| I^{m-1}(\psi(f^{m-1}(x)) - \varphi(f^{m-1}(x))) \right.$$

$$+ \left. \sum_{j=0}^{m-2} b_j I^j(\psi(f^j(x)) - \varphi(f^j(x))) \right\|$$

$$\leq \delta_2 + \left\| I^{m-1}(\psi(f^{m-1}(x)) - \varphi(f^{m-1}(x))) \right\|$$

$$+ \sum_{j=0}^{m-2} |b_j| \left\| I^j(\psi(f^j(x)) - \varphi(f^j(x))) \right\|$$

$$= \delta_2 + \|\psi(f^{m-1}(x)) - \varphi(f^{m-1}(x))\|$$

$$+ \sum_{j=0}^{m-2} |b_j| \|\psi(f^j(x)) - \varphi(f^j(x))\|$$

$$\leq \delta_2 + \left(1 + \sum_{j=0}^{m-2} |b_j|\right) M < \infty, \qquad x \in T_2.$$

Thus we have shown that (6.106) holds, which completes the proof. $\square$

Given $x \in A$, we write

$$C_f^*(x) := \{y \in A : f^n(y) = f^k(x) \text{ with some } k, n \in \mathbb{N}\},$$

$$C_f^+(x) := \{f^n(x) : n \in \mathbb{N}\},$$

$$C_f^-(x) := \{y \in A : f^n(y) = x \text{ with some } n \in \mathbb{N}\}.$$

Next, we say that $C_f^*(x)$ ($C_f^+(x)$, $C_f^-(x)$, respectively) is the orbit (positive orbit, negative orbit, resp.) of x under f. As usual, if $n \in \mathbb{N}$ and $D \subset A$, then

$$f^{-n}(D) := \{y \in A : f^n(y) \in D\}$$

and, in the case where f is injective, $x_0 \in A$ and $f^{-n}(\{x_0\}) \neq \emptyset$, we simply denote by $f^{-n}(x_0)$ the unique element of the set $f^{-n}(\{x_0\})$.

We have the following very easy observation.

Lemma 13. *Assume that* $r_1 \in \mathbb{K}$ *and* $\varphi_0 : A \to X$ *is a solution of the equation*

$$I(\varphi_0(f(x))) = r_1 \varphi_0(x). \tag{6.108}$$

Then

$$I^n(\varphi_0(f^n(x))) = r_1^n \varphi_0(x), \qquad n \in \mathbb{N}, x \in A, \tag{6.109}$$

and, in the case where f is injective and $r_1 \neq 0$,

$$\varphi_0(f^{-n}(x)) = r_1^{-n} I^n(\varphi_0(x)), \qquad n \in \mathbb{N}, x \in f^n(A). \tag{6.110}$$

Proof. Condition (6.109) follows directly from (6.108). For the proof of (6.110) it is enough to notice that (6.108) yields

$$\varphi_0(f^{-1}(x)) = r_1^{-1} I(\varphi_0(x)), \qquad x \in f(A).$$

$\square$

We need yet to prove one more auxiliary lemma. To this end let us recall that $x^* \in A$ is a non-periodic point of f provided

$$f^n(x^*) \neq x^*, \qquad n \in \mathbb{N}.$$

By A_f^* we denote the set of all non-periodic points of f.

Lemma 14. *Suppose that $S^+ \subset A_f^*$ is nonempty and such that*

$$C_f^*(x^*) \cap C_f^*(y^*) = \emptyset, \qquad x^*, y^* \in S^+, \ x^* \neq y^*, \tag{6.111}$$

I has a fixed point $u_0 \neq 0$, f is injective, $r_1 \in \mathbb{K}$, $|r_1| = 1$, $\xi : S^+ \to X$, and

$$\widehat{C}_f := \bigcup_{w^* \in S^+} C_f^-(w^*).$$

Then, for any $\delta > 0$, there is a function $\psi_0 : A \to X$, unbounded on $C_f^+(x^*)$ for every $x^* \in S^+$, such that

$$\sup_{x \in A} \|I(\psi_0(f(x))) - r_1 \psi_0(x)\| \leq \delta, \tag{6.112}$$

$$\psi_0(x) = 0, \qquad x \in S_1 := A \setminus \bigcup_{x^* \in S^+} C_f^*(x^*), \tag{6.113}$$

$$\psi_0(z^*) = \xi(z^*), \qquad z^* \in S^+, \tag{6.114}$$

$$I(\psi_0(f(x))) = r_1 \psi_0(x), \qquad x \in \widehat{C}_f \cup S_1, \tag{6.115}$$

and, for every solution $\widehat{\varphi} : A \to X$ of equation (6.105),

$$\sup_{x \in C_f^+(x^*)} \|\psi_0(x) - \widehat{\varphi}(x)\| = \infty, \qquad x^* \in S^+. \tag{6.116}$$

Proof. Take $\delta > 0$. Let $u \in X$ be a fixed point of I with $0 < \|u\| \le 1$ and $\psi_0 : A \to X$ be given by (6.113) and (6.114). Next, for each $n \in \mathbb{N}$, $x^* \in S^+$ and $w^* \in S^+$ with $f^{-n}(\{w^*\}) \ne \emptyset$, we write

$$\psi_0(f^n(x^*)) := I^{-1}(r_1\psi_0(f^{n-1}(x^*)) + r_1^n\delta u),$$

$$\psi_0(f^{-n}(w^*)) := r_1^{-1}I(\psi_0(f^{-n+1}(w^*))).$$

The definition is correct, because the sets $\widehat{C}_f$, S^+, S_1 and

$$S_2 := \bigcup_{x^* \in S^+} C_f^+(x^*),$$

are pairwise disjoint and

$$A = S_1 \cup S_2 \cup \widehat{C}_f \cup S^+.$$

Now, we prove that ψ_0 satisfies (6.112) and (6.115). To this end observe that, by (6.113), for each $x \in S_1$ we have $f(x) \in S_1$ and consequently

$$I(\psi_0(f(x))) - r_1\psi_0(x) = 0.$$

Further, in view of the definition of ψ_0 we obtain the same for every $x \in \widehat{C}_f$, which shows that (6.115) is valid.

For the proof of (6.112), take $x \in S^+ \cup S_2$. Then there are $x^* \in S^+$ and $n \in \mathbb{N}$ such that $x = f^{n-1}(x^*)$, whence the definition of ψ_0 yields

$$\|I(\psi_0(f(x))) - r_1\psi_0(x)\| = \|I(\psi_0(f^n(x^*))) - r_1\psi_0(f^{n-1}(x^*))\|$$
$$= \|r_1^n\delta u\| \le \delta.$$

Thus we have shown (6.112).

Now we show that, for each $n \in \mathbb{N}$,

$$I^n(\psi_0(f^n(x^*))) = r_1^n\psi_0(x^*) + nr_1^n\delta u, \qquad x^* \in S^+. \tag{6.117}$$

The proof is by induction on n. The case $n = 1$ is a consequence of the definition of ψ_0 for $n = 1$. So, fix an integer $n > 0$ and $x^* \in S^+$ and suppose that condition (6.117) is valid. Then, by the inductive hypothesis,

$$I^{n+1}(\psi_0(f^{n+1}(x^*))) = I^n\left(r_1\psi_0(f^n(x^*)) + r_1^{n+1}\delta u\right)$$
$$= r_1 I^n(\psi_0(f^n(x^*))) + r_1^{n+1}\delta I^n(u)$$
$$= r_1\left(r_1^n\psi_0(x^*) + nr_1^n\delta u\right) + r_1^{n+1}\delta u$$
$$= r_1^{n+1}\psi_0(x^*) + (n + 1)r_1^{n+1}\delta u.$$

This completes the proof of (6.117). Clearly, (6.117) implies that ψ_0 is unbounded on

$C_f^*(x^*)$ for each $x^* \in S^+$.

Let $\widehat{\varphi} : A \to X$ be an arbitrary solution of equation (6.105). Then, by (6.117) and Lemma 13, for every $x^* \in S^+$ and $n \in \mathbb{N}$, we have

$$I^n(\psi_0(f^n(x^*)) - \widehat{\varphi}(f^n(x^*))) = r_1^n(\psi_0(x^*) - \widehat{\varphi}(x^*)) + n r_1^n \delta u.$$

Consequently,

$$\sup_{x \in C_f^+(x^*)} \|\psi_0(x) - \widehat{\varphi}(x)\| = \infty, \qquad x^* \in S^+.$$

$\square$

Remark 31. Assume $r_j \neq 0$ for some $j \in \{1, \ldots, m\}$. Then

$$M := \{n \in \{0, \ldots, m-1\} : a_n \neq 0\} \neq \emptyset.$$

Let $\rho := \min M$. It is easily seen that (cf. Remark 30) for $\rho < m-1$, we have $b_\rho \neq 0$ and $b_j = 0$ for $j < \rho$; while for $\rho = m-1$, we get $b_j = 0$ for $j = 0, \ldots, m-2$ and $a_{m-1} = -r_1$.

Now we are in a position to prove that (under suitable assumptions) equation (6.101) is nonstable on the set S of all points of a collection of arbitrarily chosen infinite positive orbits of f in A, i.e., that there exist functions $\psi : A \to X$ satisfying inequality (6.118) (with some $\delta > 0$) and such that

$$\sup_{x \in S} \|\psi(x) - \varphi_0(x)\| = \infty$$

for each solution $\varphi_0 : A \to X$ of (6.101); from (6.121) it results that the class of such functions $\psi : A \to X$ is not small.

In the sequel, given a set $S^+ \subset A_f^*$, we write the following:

$$S_f := \bigcup_{w^* \in S^+} C_f^-(w^*),$$

$$A_f := A \setminus \bigcup_{x^* \in S^+} C_f^*(x^*), \qquad S_\rho^* := \bigcup_{i=\rho}^{m-1} f^i(S^+).$$

Proposition 13. *Let* $|r_{j_0}| = 1$ *for some* $j_0 \in \{1, \ldots, m\}$, *f be injective, $S^+ \subset A_f^*$ be nonempty and fulfills (6.111), and I have a fixed point $u \neq 0$. Then, for each $\delta > 0$*

and $\xi^* : S_\rho^* \to X$, there exists a function $\psi : A \to X$ such that

$$\sup_{x \in A} \left\| I^m(\psi(f^m(x))) + \sum_{j=0}^{m-1} a_j I^j(\psi(f^j(x))) \right\| \leq \delta, \tag{6.118}$$

$$I^m(\psi(f^m(x))) + \sum_{j=0}^{m-1} a_j I^j(\psi(f^j(x))) = 0, \qquad x \in S_f \cup A_f, \tag{6.119}$$

$$\psi(x) = 0, \qquad x \in A_f, \tag{6.120}$$

$$\psi(y) = \xi^*(y), \qquad y \in S_\rho^*, \tag{6.121}$$

and

$$\sup_{x \in C_f^+(x^*)} \|\psi(x) - \varphi_0(x)\| = \infty \tag{6.122}$$

for every $x^* \in S^+$ and each solution $\varphi_0 : A \to X$ of the equation

$$I^m(\varphi_0(f^m(x))) + \sum_{j=0}^{m-1} a_j I^j(\varphi_0(f^j(x))) = 0. \tag{6.123}$$

Proof. The case $m = 1$ follows from Lemma 14, because then $a_0 = -r_1$. So, let $m > 1$. Clearly, without loss of generality, we may assume that $j_0 = 1$.

Take $\delta > 0$. First consider the situation where $\mathbb{K} = \mathbb{C}$. To simplify the notation we assume that $\rho < m - 1$ (the case $\rho = m - 1$ is very analogous, but much simpler). Write

$$\xi(x^*) := I^{m-1}(\xi^*(f^{m-1}(x^*))) + \sum_{j=\rho}^{m-2} b_j I^j(\xi^*(f^j(x^*))), \qquad x^* \in S^+. \tag{6.124}$$

From Lemma 14 it follows that there exists a function $\psi_0 : A \to X$ such that, for every solution $\widehat{\varphi} : A \to X$ of (6.105), conditions (6.112), (6.114) and (6.116) hold.

Define $\psi : A \to X$ by (6.120), (6.121),

$$\psi(f^{\rho-n}(w^*)) := -a_\rho^{-1} I^{-\rho} \left(I^m(\psi(f^{m-n}(w^*))) + \sum_{j=\rho+1}^{m-1} a_j I^j(\psi(f^{j-n}(w^*))) \right),$$

$$w^* \in S^+, n \in \mathbb{N}, f^{\rho-n}(\{w^*\}) \neq \emptyset,$$

$$\psi(f^{m-1+n}(x^*)) := -I^{1-m}\left(\sum_{j=\rho}^{m-2} b_j I^j(\psi(f^{j+n}(x^*))) - \psi_0(f^n(x^*))\right),\qquad (6.125)$$

$$x^* \in S^+, n \in \mathbb{N}.$$

Let

$$A_1 := \{f^{m-1+n}(x^*) : x^* \in S^+, n \in \mathbb{N}\},$$

$$A_2 := \{f^{\rho-n}(w^*) : w^* \in S^+, n \in \mathbb{N}, f^{\rho-n}(\{w^*\}) \neq \emptyset\}.$$

Since the sets A_1, A_2, A_f and S_ρ^* are disjoint and

$$A = A_f \cup S_\rho^* \cup A_1 \cup A_2,$$

the definition of ψ is correct. Further, from (6.125) we deduce at once that

$$I^{m-1}(\psi(f^{m-1}(x))) + \sum_{j=0}^{m-2} b_j I^j(\psi(f^j(x))) = \psi_0(x) \qquad (6.126)$$

for every $x^* \in S^+$ and $x \in C_f^+(x^*)$.

Next, we prove (6.119). Take $x \in S_f \cup A_f$. The case $x \in A_f$ is trivial. So assume that $x \in S_f$. Then $x = f^{-n}(w^*)$ with some $w^* \in S^+$ and $n \in \mathbb{N}$ and the definition of ψ yields

$$\begin{aligned}
I^m(\psi(f^m(x))) &= I^m(\psi(f^{m-n}(w^*))) \\
&= -a_\rho I^\rho(\psi(f^{\rho-n}(w^*))) - \sum_{j=\rho+1}^{m-1} a_j I^j(\psi(f^{j-n}(w^*))) \\
&= -\sum_{j=\rho+1}^{m-1} a_j I^j(\psi(f^j(x))) - a_\rho I^\rho(\psi(f^\rho(x))) \\
&= -\sum_{j=0}^{m-1} a_j I^j(\psi(f^j(x))),
\end{aligned}$$

which completes the proof of (6.119).

Observe that, in view of (6.114), (6.121), (6.124), and Remark 31, equality (6.126) is also valid for $x \in S^+$, which means that condition (6.103) holds with $\delta_1 = \delta_2 = 0$ and $T_1 = T_2 = C_f^+(x^*) \cup \{x^*\}$ for every $x^* \in S^+$. Hence, on account of Lemma 12, for every $x^* \in S^+$, we have

$$\left\| I^m(\psi(f^m(x))) + \sum_{j=0}^{m-1} a_j I^j(\psi(f^j(x))) \right\| \leq \delta, \qquad x \in C_f^+(x^*) \cup \{x^*\}.$$

This, (6.119) and (6.120) yield (6.118). Moreover, from Lemma 12 we deduce that, for every $x^* \in S^+$, (6.122) holds for each solution $\varphi_0 : A \to X$ of equation (6.123).

It remains to consider the situation with $\mathbb{K} = \mathbb{R}$. Let $\mathbb{C}(X)$ denote the complexification of X. Define $\widehat{I} : \mathbb{C}(X) \to \mathbb{C}(X)$, $\widehat{\varphi} : A \to \mathbb{C}(X)$ and $\widehat{\xi^*} : S_\rho^* \to \mathbb{C}(X)$ by

$$\widehat{I}(x, y) := (I(x), I(y)), \qquad x, y \in X,$$

$$\widehat{\varphi}(x) := (\varphi(x), \varphi(x)), \qquad x \in A,$$

and

$$\widehat{\xi^*}(x) := (\xi^*(x), \xi^*(x)), \qquad x \in S_\rho^*.$$

By the case $\mathbb{K} = \mathbb{C}$, there is $\widehat{\psi} : A \to X^2$ such that

$$\sup_{x \in A} \left\| \widehat{I}^m(\widehat{\psi}(f^m(x))) + \sum_{j=0}^{m-1} a_j \widehat{I}^j(\widehat{\psi}(f^j(x))) \right\|_T \le \delta, \tag{6.127}$$

$$\widehat{I}^m(\widehat{\psi}(f^m(x))) + \sum_{j=0}^{m-1} a_j \widehat{I}^j(\widehat{\psi}(f^j(x))) = 0, \qquad x \in S_f \cup A_f, \tag{6.128}$$

$$\widehat{\psi}(x) = 0, \qquad x \in A_f, \tag{6.129}$$

$$\widehat{\psi}(y) = \widehat{\xi^*}(y), \qquad y \in S_\rho^*, \tag{6.130}$$

and

$$\sup_{x \in C_f^+(x^*)} \|\widehat{\psi}(x) - \widehat{\varphi}(x)\|_T = \infty$$

for every $x^* \in S^+$ and every solution $\widehat{\varphi} : A \to X^2$ of the equation

$$\widehat{I}^m(\widehat{\varphi}(f^m(x))) + \sum_{j=0}^{m-1} a_j \widehat{I}^j(\widehat{\varphi}(f^j(x))) = 0. \tag{6.131}$$

As before, let $\pi_1, \pi_2 : X^2 \to X$ be defined by: $\pi_i(x_1, x_2) := x_i$ for $x_1, x_2 \in X$, $i = 1, 2$. Fix $x^* \in S^+$ and suppose there are solutions $\varphi_1, \varphi_2 : A \to X$ of (6.123) such that

$$\sup_{x \in C_f^+(x^*)} \|\pi_1(\widehat{\psi}(x)) - \varphi_1(x)\| < \infty, \qquad \sup_{x \in C_f^+(x^*)} \|\pi_2(\widehat{\psi}(x)) - \varphi_2(x)\| < \infty.$$

Let $\widehat{\varphi}(x) := (\varphi_1(x), \varphi_2(x))$ for $x \in A$. Clearly, $\widehat{\varphi}$ is a solution to (6.131) and, by the inequalities satisfied by the Taylor norm,

$$\sup_{x \in C_f^+(x^*)} \|\widehat{\psi}(x) - \widehat{\varphi}(x)\|_T$$

$$\leq \sup_{x \in C_f^+(x^*)} \left\{ \|\pi_1(\widehat{\psi}(x)) - \pi_1(\widehat{\varphi}(x))\| + \|\pi_2(\widehat{\psi}(x)) - \pi_2(\widehat{\varphi}(x))\| \right\}$$

$$\leq \sup_{x \in C_f^+(x^*)} \|\pi_1(\widehat{\psi}(x)) - \varphi_1(x)\| + \sup_{x \in C_f^+(x^*)} \|\pi_2(\widehat{\psi}(x)) - \varphi_2(x)\| < \infty.$$

This is a contradiction.

So we deduce that, for every $x^* \in S^+$, there is $j^+(x^*) \in \{1, 2\}$ such that

$$\sup_{x \in C_f^+(x^*)} \|\pi_{j^+(x^*)}(\widehat{\psi}(x)) - \varphi_0(x)\| = \infty \tag{6.132}$$

for each solution $\varphi_0 : A \to X$ of (6.123). Now, it is enough to take

$$\psi(x) := \pi_1(\widehat{\psi}(x)), \qquad x \in A_f, \tag{6.133}$$

$$\psi(x) := \pi_{j^+(x^*)}(\widehat{\psi}(x)), \qquad x \in C_f^*(x^*), x^* \in S^+. \tag{6.134}$$

Since (6.130) yields

$$\pi_1(\widehat{\psi}(x)) = \xi^*(x) = \pi_2(\widehat{\psi}(x)), \qquad x \in S_\rho^*, \tag{6.135}$$

it is easily seen that, in view of (6.129) and (6.132), conditions (6.120)–(6.122) are valid for each solution $\varphi_0 : A \to X$ of equation (6.123) and every $x^* \in S^+$.

We have yet to prove (6.118) and (6.119). Note that, by (6.127) and (6.128), it is enough to show that, for each $x \in A$, there is $i \in \{1, 2\}$ with

$$I^m(\psi(f^m(x))) + \sum_{j=\rho}^{m-1} a_j I^j(\psi(f^j(x)))$$

$$= \pi_i \left(\widehat{I^m}(\widehat{\psi}(f^m(x))) + \sum_{j=\rho}^{m-1} a_j \widehat{I^j}(\widehat{\psi}(f^j(x))) \right). \tag{6.136}$$

Clearly, on account of (6.133), this is true for $x \in A_f$ with $i = 1$. Next, by (6.134), for each $x^* \in S^+$, (6.136) holds for $x \in C_f^*(x^*)$ with $i = j^+(x^*)$. □

Now, we show that an analogous result as in Proposition 13 can be obtained for equation (6.98). Moreover, the functions $\psi : A \to X$ satisfying the inequality

$$\sup_{x \in A} \left\| I^m(\psi_0(f^m(x))) + \sum_{j=0}^{m-1} a_j I^j(\psi_0(f^j(x))) - F(x) \right\| \leq \delta \tag{6.137}$$

(with some $\delta > 0$) and such that

$$\sup_{x \in S} \|\psi(x) - \varphi_0(x)\| = \infty$$

for a suitable set S and for each solution $\varphi_0 : A \to X$ of (6.98), can take arbitrary values at the points of the set S_ρ^* (see condition (6.140)) and be "quite close" to a solution of equation (6.98) as condition (6.138) shows, and even to a given solution of the equation (see (6.139)).

Theorem 117. *Assume that I has a fixed point $u \neq 0$, $|r_{j_0}| = 1$ for some $j_0 \in \{1, \ldots, m\}$, f is injective, $S^+ \subset A_f^*$ is nonempty and fulfills (6.111), $\xi_0^* : S_\rho^* \to X$, and there is a solution $\eta : A \to X$ to equation (6.98). Then, for each $\delta > 0$, there exists a function $\psi_0 : A \to X$ such that (6.137) holds,*

$$I^m(\psi_0(f^m(x))) + \sum_{j=0}^{m-1} a_j I^j(\psi_0(f^j(x))) = F(x), \qquad x \in S_f \cup A_f, \tag{6.138}$$

$$\psi_0(x) = \eta(x), \qquad x \in A_f, \tag{6.139}$$

$$\psi_0(y) = \xi_0^*(y), \qquad y \in S_\rho^*, \tag{6.140}$$

and

$$\sup_{x \in C_f^+(x^*)} \|\psi_0(x) - \varphi(x)\| = \infty \tag{6.141}$$

for each solution $\varphi : A \to X$ of equation (6.98) and every $x^* \in S^+$.

Proof. Take $\delta > 0$. Let $\xi^* : S_\rho^* \to X$ be given by

$$\xi^*(y) = \xi_0^*(y) - \eta(y), \qquad y \in S_\rho^*.$$

Then, by Proposition 13, there is a function $\psi : A \to X$ such that conditions (6.118)–(6.122) are fulfilled for every $x^* \in S^+$ and each solution $\varphi_0 : A \to X$ of equation (6.123). Define $\psi_0 : A \to X$ by

$$\psi_0(x) = \psi(x) + \eta(x), \qquad x \in A.$$

Then it is easily seen that (6.118)–(6.121) imply (6.137)–(6.140).

Next, let $\varphi : A \to X$ be a solution of equation (6.98). Then, $\varphi_0 := \varphi - \eta$ is a solution to (6.123), whence (6.122) holds for every $x^* \in S^+$. Consequently,

$$\sup_{x \in C_f^+(x^*)} \|\psi_0(x) - \varphi(x)\| = \sup_{x \in C_f^+(x^*)} \|\psi(x) - \varphi_0(x)\| = \infty, \qquad x^* \in S^+.$$

This completes the proof. $\square$

If we take $I(x) = x$ for every $x \in X$ in Theorem 117, then we obtain at once the following slightly weaker version of the main result in [7].

Corollary 18. *Let* $|r_{j_0}| = 1$ *for some* $j_0 \in \{1, \ldots, m\}$, f *be injective,* $S^+ \subset A_f^*$ *be nonempty and fulfill* (6.111), *and* $\xi^* : S_\rho^* \to X$. *Suppose that* $\eta : A \to X$ *is a solution of the equation*

$$\eta(f^m(x)) + \sum_{j=0}^{m-1} a_j \eta(f^j(x)) = F(x). \tag{6.142}$$

Then, for each $\delta > 0$, *there exists a function* $\psi_0 : A \to X$ *such that*

$$\sup_{x \in A} \left\| \psi_0(f^m(x)) + \sum_{j=0}^{m-1} a_j \psi_0(f^j(x)) - F(x) \right\| \leq \delta, \tag{6.143}$$

$$\psi_0(f^m(x)) + \sum_{j=0}^{m-1} a_j \psi_0(f^j(x)) = F(x), \qquad x \in S_f \cup A_f, \tag{6.144}$$

and conditions (6.139)–(6.141) *are satisfied for every* $x^* \in S^+$ *and each solution* $\varphi : A \to X$ *of* (6.142).

Remark 32. One of the assumptions of Theorem 117 is that (6.98) admits a solution $\varphi \in X^A$. In the case where

$$A = \bigcup_{x^* \in A_f^*} C_f^*(x^*)$$

such a solution can be constructed analogously as in the proof of Theorem 117, by taking any values for $\varphi(f^i(x^*))$, with $i = \rho, \ldots, m-1$ and $x^* \in A_f^*$, and defining further φ by

$$\varphi(f^{m+n-1}(x^*)) := I^{-m} \left(- \sum_{j=\rho}^{m-1} a_j I^j(\varphi(f^{j+n-1}(x^*))) + F(f^{n-1}(x^*)) \right),$$

$$n \in \mathbb{N},$$

$$\varphi(f^{\rho-n}(x^*)) := -I^{-\rho} \left(a_\rho^{-1} \left(I^m(\varphi(f^{m-n}(x^*))) + \sum_{j=\rho+1}^{m-1} a_j I^j(\varphi(f^{j-n}(x^*))) \right. \right.$$

$$\left. \left. -F(f^{-n}(x^*)) \right) \right), \qquad n \in \mathbb{N}, f^{\rho-n}(\{x^*\}) \neq \emptyset,$$

where ρ is as in Remark 31.

Next, observe that if (6.98) does not have any solution in X^A and F is bounded, then (6.98) is Ulam nonstable in X^A, because in such a case every bounded function $\psi_0 \in X^A$ satisfies (6.137).

The assumption of injectivity of f plays a crucial role in the proof of Proposition 13, and therefore, it seems to be important for Theorem 117, as well; but actually nonstability of (6.98) in X^A also can be obtained under an assumption somewhat weaker than injectivity of f or even without any such assumption. The next corollary and a very simple theorem show that.

Corollary 19. *Assume that $|r_j| = 1$ for some $j \in \{1, \ldots, m\}$, equation (6.98) has a solution $\varphi \in X^A$, and there is $x^* \in A_f^*$ such that f is injective on the set $C_f^*(x^*)$, i.e.,*

$$f(x) \neq f(y), \qquad x, y \in C_f^*(x^*), \ x \neq y.$$

Then equation (6.98) is Ulam nonstable in X^A.

Proof. According to Theorem 13 (with $A := C_f^*(x^*)$ and $S^+ := \{x^*\}$) there exists a function $\psi : C_f^*(x^*) \to X$ such that

$$\sup_{x \in C_f^*(x^*)} \left\| I^m(\psi(f^m(x))) + \sum_{j=0}^{m-1} a_j I^j(\psi(f^j(x))) - F(x) \right\| =: \delta < \infty$$

and

$$\sup_{x \in C_f^+(x^*)} \|\psi(x) - \varphi_0(x)\| = \infty$$

for each solution $\varphi_0 : C_f^*(x^*) \to X$ of equation (6.98). Write

$$\psi(x) := \varphi(x), \qquad x \in A \setminus C_f^*(x^*).$$

Since φ is a solution of (6.98) and

$$f(A \setminus C_f^*(x^*)) \cap C_f^*(x^*) = \emptyset,$$

we have

$$\sup_{x \in A} \left\| I^m(\psi(f^m(x)) + \sum_{j=0}^{m-1} a_j I^j(\psi(f^j(x))) - F(x) \right\|$$

$$= \sup_{x \in C_f^*(x^*)} \left\| I^m(\psi(f^m(x))) + \sum_{j=0}^{m-1} a_j I^j(\psi(f^j(x))) - F(x) \right\| = \delta.$$

But, on the other hand, for each solution $\varphi_0 : A \to X$ to (6.98),

$$\sup_{x \in A} \|\psi(x) - \varphi_0(x)\| \geq \sup_{x \in C_f^+(x^*)} \|\psi(x) - \varphi_0(x)\| = \infty.$$

This completes the proof. $\qquad\qquad\qquad\qquad\qquad\qquad\qquad\qquad\qquad\qquad\qquad$ □

Theorem 118. *Let $m > 1$, $S \subset A$ be nonempty, $f(S) \subset S$,*

$$\sup_{x \in S} \|F(x)\| < \infty, \tag{6.145}$$

$$\lim_{n \to \infty} \left\| \sum_{k=0}^{n} I^k(F(f^k(x_0))) \right\| = \infty \tag{6.146}$$

for some $x_0 \in S$, and

$$\sum_{j=0}^{m-1} a_j + 1 = 0. \tag{6.147}$$

Then equation (6.98) is Ulam nonstable on S in X^A.

Proof. Clearly, (6.147) implies that $r_j = 1$ for some $j \in \{1, \ldots, m\}$. Without loss of generality we may assume that $j = 1$. In view of (6.145), the function $\psi : A \to X$, $\psi(x) = 0$ for $x \in A$, satisfies

$$\sup_{x \in S} \left\| I^m(\psi(f^{m-1}(x))) + \sum_{j=0}^{m-1} a_j I^j(\psi(f^j(x))) - F(x) \right\| < \infty. \tag{6.148}$$

For the proof by contradiction suppose that there is a solution $\varphi : A \to X$ to (6.98) such that

$$\sup_{x \in S} \|\psi(x) - \varphi(x)\| =: M < \infty \tag{6.149}$$

and write

$$\widehat{\eta}(x) := I^{m-1}(\varphi(f^{m-1}(x))) + \sum_{j=0}^{m-2} b_j I^j(\varphi(f^j(x))), \qquad x \in A. \tag{6.150}$$

Then $\widehat{\eta} : A \to X$ satisfies the equation

$$I(\widehat{\eta}(f(x))) = r_1 \widehat{\eta}(x) + F(x) \tag{6.151}$$

and (6.149) yields

$$\|\psi(x) - \widehat{\eta}(x)\| = \left\|I^{m-1}(\psi(f^{m-1}(x)) - \varphi(f^{m-1}(x)))\right.$$

$$\left. + \sum_{j=0}^{m-2} b_j I^j(\psi(f^j(x)) - \varphi(f^j(x)))\right\|$$

$$\leq \left\|I^{m-1}(\psi(f^{m-1}(x))) - \varphi(f^{m-1}(x)))\right\|$$

$$+ \sum_{j=0}^{m-2} |b_j| \left\|I^j(\psi(f^j(x)) - \varphi(f^j(x)))\right\|$$

$$= \|\psi(f^{m-1}(x))) - \varphi(f^{m-1}(x))\|$$

$$+ \sum_{j=0}^{m-2} |b_j| \|\psi(f^j(x)) - \varphi(f^j(x))\|$$

$$\leq \left(1 + \sum_{j=0}^{m-2} |b_j|\right) M < \infty, \qquad x \in S.$$

On the other hand, by (6.146), $\widehat{\eta}: A \to X$ is unbounded on S, because

$$I^n(\widehat{\eta}(f^n(x_0))) = \widehat{\eta}(x_0) + \sum_{k=0}^{n-1} I^k(F(f^k(x_0))), \qquad n \in \mathbb{N}.$$

Consequently,

$$\sup_{x \in S} \|\psi(x) - \widehat{\eta}(x)\| = \sup_{x \in S} \|\widehat{\eta}(x)\| = \infty. \tag{6.152}$$

This contradiction completes the proof.   □

REFERENCES

1. J.A. Baker, The stability of certain functional equations, Proc. Amer. Math. Soc. 112 (1991) 729–732.
2. J. Brzdęk, Remarks on stability of some inhomogeneous functional equations, Aequationes Math. 89 (2015) 83–96.
3. J. Brzdęk, D. Popa, B. Xu, Note on nonstability of the linear recurrence, Abh. Math. Sem. Univ. Hamburg 76 (2006) 183–189.
4. J. Brzdęk, D. Popa, B. Xu, The Hyers-Ulam stability of nonlinear recurrences, J. Math. Anal. Appl. 335 (2007) 443–449.
5. J. Brzdęk, D. Popa, B. Xu, On nonstability of the linear recurrence of order one, J. Math. Anal. Appl. 367 (2010) 146–153.
6. J. Brzdęk, D. Popa, B. Xu, Remarks on stability of the linear recurrence of higher order, Appl. Math. Lett. 23 (2010) 1459–1463.
7. J. Brzdęk, D. Popa, B. Xu, Note on nonstability of the linear functional equation of higher order, Comput. Math. Appl. 62 (2011) 2648–2657.
8. J. Brzdęk, D. Popa, B. Xu, Remarks on stability and non-stability of the linear functional equation of the first order, Appl. Math. Comput. 238 (2014) 141–148.
9. Z. Gajda, On stability of additive mappings, Int. J. Math. Math. Sci. 14 (1991) 431–434.

10. D.H. Hyers, On the stability of the linear functional equation, Proc. Nat. Acad. Sci. USA 27 (1941) 222–224.
11. D.H. Hyers, G. Isac, Th.M. Rassias, Stability of Functional Equations in Several Variables, Birkhäuser, Boston - Basel - Berlin, 1998.
12. S.M. Jung, On a general Hyers-Ulam stability of gamma functional equation, Bull. Korean Math. Soc. 34 (1997) 437–446.
13. S.M. Jung, On the modified Hyers-Ulam-Rassias stability of the functional equation for gamma function, Mathematica 39 (62) (1997) 233–237.
14. S.M. Jung, On the stability of gamma functional equation, Results Math. 33 (1998) 306–309.
15. G.H. Kim, On the stability of generalized gamma functional equation, Int. J. Math. Math. Sci. 23 (2000) 513–520.
16. G.H. Kim, B. Xu, W.N. Zhang, Notes on stability of the generalized gamma functional equation, Int. J. Math. Math. Sci. 32 (2002) 57-63.
17. M. Kuczma, B. Choczewski, R. Ger, Iterative Functional Equations, Encyclopedia of Mathematics and its Applications, Cambridge University Press, 1990.
18. S.H. Lee, K.W. Jun, The stability of the equation $f(x + p) = kf(x)$, Bull. Korean Math. Soc. 35 (1998) 653–658.
19. D. Popa, Hyers-Ulam-Rassias stability of a linear recurrence, J. Math. Anal. Appl. 309 (2005) 591–597.
20. D. Popa, Hyers-Ulam stability of the linear recurrence with constant coefficients, Adv. Difference Equ. 2005 (2005) 101–107.
21. T. Trif, On the stability of a general gamma-type functional equation, Publ. Math. Debrecen 60 (2002) 47–61.

INDEX

Printed in the United States
By Bookmasters